PROTEIN MISFOLDING

PROTEIN MISFOLDING

CIAN B. O'DOHERTY AND ADAM C. BYRNE
EDITORS

Nova Biomedical Books
New York

For permission to use material from this book please contact us:
Telephone 631-231-7269; Fax 631-231-8175
Web Site: http://www.novapublishers.com

LIBRARY OF CONGRESS CATALOGING-IN-PUBLICATION DATA

Protein misfolding / Cian B. O'Doherty and Adam C. Byrne (editors).
p. ; cm.
Includes index.
ISBN 978-1-60456-881-3 (hardcover)
1. Protein folding. 2. Amyloid. I. O'Doherty, Cian B. II. Byrne, Adam C.
[DNLM: 1. Protein Folding. 2. Protein Conformation. QU 55.9 P9663 2008]
QP551.P697222 2008
572'.633--dc22

2008025039

Published by Nova Science Publishers, Inc. ✢ New York

CONTENTS

PREFACE

Proteins occasionally exhibit the potential to adopt alternate folds under natural or artificial conditions. These conformational changes can lead to the creation of disease-forming amyloid plaques or accumulation of insoluble aggregates in over-expressed recombinant proteins. Because alternate folds challenge the traditional notion of structurally defined proteins and frequently lead to problems in natural and laboratory settings, it is tempting to characterize structurally ambiguous proteins as ill-behaved. Unfortunately, this perspective masks the positive role structural ambiguity can play in protein function and evolution. This book provides leading-edge research on the study of protein misfolding from researchers all over the globe.

Chapter I - Amyloidoses are protein misfolding diseases characterized by fibrillar protein deposits. The complexity and conformational heterogeneity of these aggregates makes their structural characterization a cutting edge problem in modern biophysics. Solution and solid-state nuclear magnetic resonance (NMR) have been at the forefront of amyloid structural biology. The authors review recent progress in NMR investigations of amyloidogenic proteins with a particular emphasis on amyloid beta - the peptide constituent of fibrils found in patients with Alzheimer's disease, and alpha-synuclein - the protein most closely linked with Parkinson's disease. The authors bvegin by highlighting solution NMR work on the structures and dynamics of the initial monomeric states. NMR investigations of micelle bound states are summarized as they give insights into the interactions of amyloidogenic proteins with membranes, associations that are likely to be important in both normal and pathological function. The authors describe results from indirect NMR studies of fibrils that use a quenched hydrogen exchange method to transfer structural information from aggregates to the dissociated monomeric state. Finally, the authors review recent progress in solid-state NMR studies that are beginning to provide detailed structural models of amyloid fibrils. Throughout this chapter the authors have tried to highlight the shared properties of amyloidogenic proteins and to describe the unifying themes emerging in mechanistic models of fibrillar protein misfolding.

Chapter II - Proteins occasionally exhibit the potential to adopt alternate folds under natural or artificial conditions. These conformational changes can lead to the creation of disease-forming amyloid plaques or accumulation of insoluble aggregates in over-expressed recombinant proteins. Because alternate folds challenge the traditional notion of structurally defined proteins and frequently lead to problems in natural and laboratory settings, it is tempting to characterize structurally ambiguous proteins as ill-behaved. Unfortunately, this perspective masks the positive role structural ambiguity can play in protein function and evolution. With an estimated 30% of eukaryotic proteins containing extensive disordered regions and mutagenesis experiments showing that a minimal number of polymorphisms are required to specify dramatically different structures, protein instability and conformational switching may play a greater role in cell biology than previously expected. The authors review an expanding body of research that encourages a more relaxed interpretation of the central dogma and promotes a model for protein evolution in which conformational switching uncovers novel function by exposing cryptic epitopes. Under this model, prion-induced diseases represent a negative outcome of a natural process that helps organisms explore available sequence space for positive biological activity.

Chapter III - Protein folding remains one of the most puzzling problems in molecular biology. Protein folding is now widely assumed to be co-translational occurring as the polypeptide chain is being synthesized on the ribosome. With recognition of the co-translational nature of *in vivo* protein folding, the focus of many recent studies has shifted from the investigation of protein folding in a test tube (from the unfolded state) to the investigation of protein folding in the cell or in cell-free translation systems. Interest in the problem of *in vivo* protein folding has increased tremendously in recent years with the broad recognition of the central role of protein misfolding in many degenerative diseases and pathological conditions as well as with the increased need to produce correctly folded recombinant and therapeutic proteins. Numerous quality-control mechanisms exist in vivo which aid identification and removal of abnormally folded and misfolded proteins. Cells evolved two major quality control systems to deal with misfolded proteins: molecular chaperones which help to stabilize proteins and bring them back to the productive folding pathway, and the ubiquitin proteasome system, which degrades misfolded proteins that can not be refolded to the native state. The phenomenon of co-translational ubiquitination (and degradation) of proteins suggests that misfolding of proteins can start very early during protein synthesis on the ribosome. What are the factors influencing the ability of proteins to fold correctly during the process of translation? A number of experimental data and recent *in silico* modeling experiments suggest that the rate at which various portions of the growing polypeptide chain emerge from the ribosome can influence both the folding pathway and the final conformation of the protein. It has been thus suggested that abnormal translation rates may lead to increased levels of protein misfolding. This review summarizes recent developments and the prevalent literature covering this topic.

Chapter IV - Hereditary hemochromatosis (HH) refers to a unique clinicopathologic subset of iron overload syndromes that include the disorder related to the C282Y homozygote mutation of the hemochromatosis protein HFE (the most common form of hemochomatosis). HFE C282Y is a mutant protein the forms aggregates and is subsequently retained in the Endoplasmic Reticulum (ER). The aberrant accrual of proteins can therefore lead to ER stress which may have detrimental effects on normal intracellular function contributing to liver pathology. Recent advances are shedding light on the potential role of HFE C282Y in eliciting pathogenesis via ER stress. All available data are pointing towards a new paradigm that of, HFE C282Y HH as a conformational disorder. Conformational diseases are a class of disorders associated with an underlying accrual of mutant protein and examples include conditions such as Alzheimer's, Parkinson's, Huntington's diseases and the liver disease Z alpha 1-antitrypsin deficiency. This review sets out to examine this new paradigm for HFE C282Y HH liver pathogenesis with an emphasis on the consequences of HFE C282Y protein retention in the ER. Viewing HH from the perspective of these unified conditions will enable greater insight into the molecular events underlying this disease and provides a new vista on this metabolic iron overload disease. Furthermore, the recognition of HFE C282Y HH from this perspective suggests intriguing possibilities for new and more rational approaches to the diagnosis and treatment for this liver disease.

Chapter V - In addition to "normal," native protein structure, some proteins can also form alternative, misfolded structures. During the past years, it has been shown that some diseases are connected with protein misfolding and the formation of insoluble aggregates called amyloid plaques. For some proteins which are capable to form amyloid structures, those regions which are important for amyloid formation are already experimentally outlined (they are called amyloidogenic regions). In these proteins, the authors predicted those residues that are important for "normal" folding (that is, which are involved into the folding nucleus of the native structure) and compared them with amyloidogenic ones. The average of the predicted Φ-values (which reflect the degree of involvement of the amino acid residue into the folding nucleus) over 12 amyloidogenic regions (of 7 globular proteins in which amyloidogenic regions are now localized experimentally) is significantly greater than the average Φ-value averaged over residues outside amyloidogenic regions. This demonstrates that amino acid residues in amyloidogenic regions in average are more included into folding nucleus than amino acid residues from non-amyloidogenic regions. In total, 8 of 12 amyloidogenic fragments are located in the folding nuclei. This is an indication that amyloidogenic regions are typically incorporated into the native structure early during its formation (not later than at the rate-limiting step of a "normal" folding process).

Chapter VI - Many aspects regulating the machineries of protein folding and degradation in the yeast *S. cerevisiae* are conserved in mammalian cells. Due to the lower redundancy and to the facility of manipulating the yeast genome, many players involved in protein folding in the endoplasmic reticulum (ER) and in degradation of terminally misfolded proteins from the ER lumen (ERAD) have been first discovered in the yeast *S. cerevisiae*. *S. cerevisiae* contains simplified versions of the corresponding mammalian pathways and probably all yeast players have functional orthologs in mammalian cells. However, metazoans certainly benefit from many additional components that are still poorly characterized.

Here the authors highlight differences and similarities between *S. cerevisiae* and mammalian polypeptides in the exploitation of the calnexin chaperone system during glycoprotein maturation and the intervention of a recently described ER chaperone, Yos9p/OS-9, when polypeptide maturation has failed.

Chapter VII - The endoplasmic reticulum (ER) stress response has been recognized as a cytoprotective mechanism to cope with unfolded proteins accumulated in the ER, although the exact significance of the ER stress response in biological processes remains elusive. In this chapter, the author will try to redefine the ER stress response as an autoregulatory mechanism that controls the amount of the ER in accordance with cellular demand. Moreover, since recent discoveries suggest that mechanisms regulating the amount or capacity of other organelles including the mitochondria, peroxisomes and Golgi apparatus also exist in eukaryotic cells, the author proposes the working hypothesis that the amount of eukaryotic organelles is automatically regulated by signaling between the nucleus and each organelle, to adjust each organelle's capacity according to cellular demand.

Chapter VIII - Neurodegenerative disorders such as Amyotrophic lateral sclerosis (ALS), Alzheimer's disease, Parkinson's disease, Prion-related disorders (PrD) and Huntington's disease (HD) share a common neuropathology, primarily featuring the presence of abnormal protein inclusions containing specific misfolded proteins. Recent evidence indicates that alteration in organelle function is a common pathological feature of protein misfolding disorders. The endoplasmic reticulum (ER) is an essential compartment for protein folding, maturation, and secretion. Signs of ER stress have been extensively described in most experimental models of neurological disorders and more recently in the brains of human patients. ER stress is caused by functional disturbances which result in accumulation of unfolded/misfolded proteins at the ER lumen. To cope with ER stress, cells activate an integrated signaling response termed the Unfolded Protein Response (UPR) which aims to reestablish homeostasis through transcriptional upregulation of genes involved in folding, protein quality control and protein degradation pathways including lysosome-mediated autophagy. Small molecules with chaperone-like activity have been shown to alleviate ER stress and decrease protein misfolding in disease settings. In this chapter the authors give an overview of the role of ER stress in pathological conditions such as protein misfolding disorders and brain ischemia, and discuss possible pharmacological strategies to target the UPR with therapeutic benefits.

Chapter IX - The authors have recently discovered that 20 amino acid long sequence fragments of natural proteins, pair-wise similar to one another, can be organized in "walks" and networks, such that members of the same network may have rather different sequences, yet are related either functionally or structurally, or both. Traditional sequence comparisons would not classify such dissimilar sequence fragments, taken separately, as belonging to the same family. The sequence relatedness via many intermediates, i.e., via connections within the network in the sequence space, is the next, more advanced step in the development of sequence comparison methods. The networks suggest the kinship even between seemingly unrelated sequences, often belonging to different protein families.

The use of short fragments for the sequence comparisons reveals that some segments of a given protein sequence belong to large networks of the sequence space, i.e., have many close and distant relatives across species. At the same time, other segments are rather poorly represented in the space. As a result, the most informative sequence segments of a protein are identified, revealing a modular structure of the proteins.

Remarkably, such mapping by networks reveals existence of "silent" regions in the protein sequences. These regions belong to a very small networks, often of only one point size. These singular sequence fragments would be antonyms of conserved sequences, which appear in many proteomes. The first report of investigation of the nature of such silent regions in the protein sequences will be described in the chapter.

Short Communication A - Small-angle scattering (SAS) of X-rays (SAXS) and neutrons (SANS), have been widely used to study the overall structure of folded and unfolded proteins in solution. SAS measurements of unstructured states are normally analysed in terms of radii of gyration (R_g) that are either monitored during protein folding/unfolding processes or compared with those predicted for native states or random coil models. The authors' groups have recently developed a new methodology, the so-called Ensemble Optimization Method (EOM), analysing the scattering data in terms of ensembles of coexisting conformations, which, when treated collectively fit the experimental profile. This strategy allows the description of size properties of disordered proteins in solution as distributions within their ensembles. Therefore, the traditional single value averaged R_g description of a flexible state can be converted into a more detaied picture that is the density of accessible R_g values. Thanks to this, even short fragments containing secondary structure, or presence of transient interactions between remote regions of the chain can sometimes be identified. The localization of these structural features can be addressed through the simultaneous analysis of profiles measured for different fragments (e.g. deletion mutants) of the protein.

Short Communication B - Amyloid fibrils and plaques that form deposits are characteristic of several degenerative conditions affecting a large variety of tissues. Almost constantly new evidences are linking protein misfolding and aggregation to one or more type of diseases, leading to the hypothesis that a considerable part of the known diseases are caused by the increase of these non-functional molecules. Molecular chaperones are a family of proteins initially identified as heat shock proteins that are involved with assistance of the correct folding and prevention of aggregation of unfolded polypeptides. The constitutive induction and the housekeeping characteristics of these proteins were recognized later. In addition to that, recent works have shown that these proteins are also capable of disaggregation and refolding of stable protein aggregates, a novel function so spectacular that raised the hypothesis that molecular chaperones may be as important in regulating the malefic effects of misfolded or aggregated proteins as in raising the yield of functional molecules. These new findings have also led to the hypothesis that the functional failure of the molecular chaperone machinery is a potential trigger of degenerative conditions. A functional failure that would be more common as the organism ages favoring degenerative diseases in the elderly. In conclusion, new evidences on the function of molecular chaperones have enhanced the potential of these proteins as targets for therapeutic approaches aimed in decreasing the toxicity of aggregates.

In: Protein Misfolding
Editors: C. B. O'Doherty, A.C. Byrne

ISBN: 978-1-60456-881-3

Chapter I

NMR OF AMYLOIDOGENIC PROTEINS

Andrei T. Alexandrescu* and Robyn L. Croke
Department of Molecular and Cell Biology, University of Connecticut
Storrs, CT 06269-3125, U.S.A.

ABSTRACT

Amyloidoses are protein misfolding diseases characterized by fibrillar protein deposits. The complexity and conformational heterogeneity of these aggregates makes their structural characterization a cutting edge problem in modern biophysics. Solution and solid-state nuclear magnetic resonance (NMR) have been at the forefront of amyloid structural biology. We review recent progress in NMR investigations of amyloidogenic proteins with a particular emphasis on amyloid beta - the peptide constituent of fibrils found in patients with Alzheimer's disease, and alpha-synuclein - the protein most closely linked with Parkinson's disease. We begin by highlighting solution NMR work on the structures and dynamics of the initial monomeric states. NMR investigations of micelle bound states are summarized as they give insights into the interactions of amyloidogenic proteins with membranes, associations that are likely to be important in both normal and pathological function. We describe results from indirect NMR studies of fibrils that use a quenched hydrogen exchange method to transfer structural information from aggregates to the dissociated monomeric state. Finally, we review recent progress in solid-state NMR studies that are beginning to provide detailed structural models of amyloid fibrils. Throughout this chapter we have tried to highlight the shared properties of amyloidogenic proteins and to describe the unifying themes emerging in mechanistic models of fibrillar protein misfolding.

* Correspondence concerning this article should be addressed to: Andrei Alexandrescu, Department of Molecular and Cell Biology, University of Connecticut, 91 North Eagleville Road, U-3125, Storrs, CT 06269-3125, U.S.A. Tel:(860)-486-4414; Fax: (860)-486-4331; e-mail andrei@uconn.edu.

Abbreviations

Aβ	(Alzheimer) amyloid β-peptide
AD	Alzheimer Disease
AFM	atomic force microscopy
APP	(Alzheimer) amyloid precursor protein
αS	α-synuclein
DMSO	dimethyl sulfoxide
EM	electron microscopy
IAPP	islet amyloid polypeptide
GAG	glycosaminoglycan
HSPG	heparan sulfate proteoglycan
β2m	beta-2 microglobulin
HSQC	heteronuclear single quantum coherence
HX	hydrogen exchange
NMR	nuclear magnetic resonance
NOE	nuclear Overhauser enhancement
PRE	paramagnetic relaxation enhancement
PrP^C	soluble form of prion protein
PrP^{Sc}	protease-resistant form of prion protein
qHX	quenched hydrogen exchange
RDC	residual dipolar coupling
SAA	serum amyloid A
SDS	sodium dodecyl sulfate
ssNMR	solid state NMR

1. Introduction

Amyloid diseases are associated with some 22 different types of proteins that acquire aberrant function over time and assemble into fibrillar deposits [1-3]. The fibrils exhibit similar morphologies by EM (Figure 1) and AFM; have a characteristic cross-β-sheet X-ray fiber diffraction pattern, and similar tinctorial properties [3-5]. A large variety of unrelated proteins including bacterial proteins [6], proteins with native α-helical structure [7], and designed peptides as small as six amino acids [8] are able to form fibrils with superstructural properties similar to the fibrils formed from disease-causing proteins. This has led to the hypothesis that amyloid formation could be a general property of proteins [9]. Of the proteins involved in amyloid disease, some have stable native tertiary structures such as lysozyme [10], transthyretin [11], prion protein [12], and β-microglobulin [13]. Others appear to be "intrinsically unfolded" in isolation. These include α-synuclein (αS) [14], amyloid beta peptide (Aβ) [15], IAPP [16], and calcitonin [17].

A common mechanistic theme amongst amyloidogenic proteins is that aggregation proceeds through partially folded intermediates [2]. Such intermediates can be generated either through partial unfolding of globular structures, or through partial folding of intrinsically unfolded precursors [1]. At this time, it has not been conclusively established whether the toxicity of amyloidogenic proteins is due to the fibrils themselves or due to smaller soluble aggregates that precede or derive from the fibrils [3,5]. Evidence linking plaques to cytotoxicity include amyloid "infectivity" in a mouse model, where pre-existing AD plaques from one animal were shown to seed plaque formation when injected into another [18]. The formation of small AD plaques has also been recently shown to precede neuron dystrophy in an elegant multiphoton fluorescence microscopy study [19]. As pointed out by the authors, however, these findings could also be consistent with the plaques constituting a reservoir of dissociable toxic oligomers [18,19]. Indeed, the consensus emerging in the field is that soluble aggregates are more likely to be responsible for toxicity [1,20-23]. Evidence supporting this view includes that the number and sizes of fibrillar deposits identified in post-mortem samples correlates only weakly with the severity of symptoms, that plaques can be found in asymptomatic apparently normal individuals, and that some subtypes of diseases involving amyloidogenic proteins occur without plaques [21-23]. If soluble oligomers are indeed the culprits, the insoluble fibrils could be a side-effect, may confer protection against the more toxic soluble aggregates, or could serve as a stable reservoir of short-lived soluble aggregates. Extrinsic factors that interact with amyloidogenic proteins such as proteoglycans [24], metals [25], metabolites [26], or cell membranes [27] may also have roles in the disease processes.

Because of the breadth and complexity of the amyloid field our review will focus primarily on the two proteins αS and Aβ which are responsible for the most prevalent diseases: Parkinson's disease (PD) and Alzheimer's disease (AD), respectively. According to the World Health Organization statistics, there are 18 million AD and 4 million PD sufferers worldwide. Over the next 20 years these numbers are expected to double as human life spans increase, and as the population of developed countries becomes older. In addition to the toll in human suffering, the burdens placed on society will be enormous; the cost of treating AD in the US alone is currently over 100 billion dollars a year [28]. Understanding the mechanisms of these diseases is therefore a major challenge for modern structural biology.

Alzheimer's disease is characterized by brain deposits of a family of 39-43 residue peptides, whose tendency to form amyloid increases with peptide length. The peptides result from cleavage of a transmembrane α-helix segment of amyloid precursor protein (APP) by β- and γ-secretase under the influence of the protein presenilin [5]. Aβ peptides are found in a variety of tissues, yet fibrillar deposits are specific to synapses and the basement membranes of brain blood vessels [29,30]. Mutations in the APP and presenilin genes predispose people to Alzheimer's, demonstrating a genetic link between the amyloidogenic peptides and the disease [31]. The Aβ peptides are innocuous in their monomeric unfolded forms but have been demonstrated to become neurotoxic when aggregated [18,32,33].

Alpha-synuclein (αS) is a protein expressed predominantly at the pre-synaptic terminals of brain neurons [34]. The physiological functions of αS are unknown, although synaptic vesicle recycling is thought to be a likely role [34-36]. Genetic and pathology evidence implicates αS as the principal culprit in the etiology of PD. The αS missense mutations A53T, A30P, and E46K, as well as a triplication of the wild-type gene result in hereditary diseases that include Parkinsonism together with more widespread symptoms [34-37]. In its fibrillar form, αS is the principal component of Lewy body deposits, the neuropathological hallmarks of PD [34,35,37,38]. In contrast to extracellular amyloids, Lewy bodies are intracellular, occurring in the dopaminergic neurons of the brain stem's *substantia nigra* [39]. αS is also found in a number of other Lewy body pathologies [34,35], is secreted outside of cells [40], and an extracellular 35-residue fragment of αS occurs as the non-Aβ-component (NAC) of AD plaques [41].

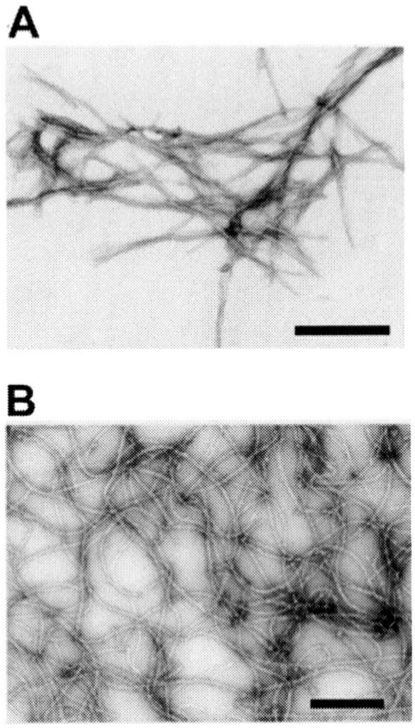

Figure 1. Electron micrographs of (A) Aβ(1-40) and (B) αS. Fibrils were negatively stained with 1% uranyl acetate. The bars represent 100 nm.

Current mechanistic pictures suggest an assortment of conformations and conformational transitions are important in the biology of amyloidogenic proteins. These include the native or nascent forms of the proteins, membrane-bound species, soluble oligomers, and insoluble fibrils. We will describe NMR investigations of these classes of states in subsequent sections of this chapter. The strengths of NMR spectroscopy for studies of amyloidogenic proteins include an unmatched ability to provide structural information on conformational transitions at atomic resolution [42]. Amongst other applications this can be used to characterize binding reactions - for example the association of amyloidogenic proteins with each other, with membranes, or with small molecules that either promote or hinder aggregation. The ability to study proteins in solution makes it possible to see how changes in solution conditions affect the conformational properties of amyloidogenic proteins. It is thus relatively simple to detect changes in populations of states due to changes in pH, salt, temperature, or time. The exquisite sensitivity of NMR to disorder and flexibility has contributed greatly to our understanding of unfolded and partially folded states [43], which in turn has provided valuable insight into the mechanisms by which such states misfold into toxic aggregates. The principal limitation of solution NMR, is that aggregates and complexes with molecular weights above 20-30 kDa are exceedingly difficult to study, although advances in the technique continue to push back the size limit of NMR [42]. In some cases the size limitation of NMR is useful as a molecular weight filter, however, amyloid aggregates and particularly fibrils cannot be detected by standard solution NMR methods. The problem can be circumvented to an extent by indirect quenched hydrogen exchange methods that translate information from the invisible aggregates to monomeric forms that can be studied by solution NMR [44]. More recently, solid state NMR (ssNMR) has developed to the point where it is starting to provide structures with detail comparable to solution NMR [45,46]. Because the fibrillar forms of amyloidogenic proteins are in the solid state yet lack the three-dimensional order of crystals, ssNMR is uniquely suited for the structural studies of these aggregates. Indeed, the most detailed structural models of amyloid fibrils to date have come from ssNMR [47-49]. Excitingly, the ability to freeze fibrillogenic intermediates in the solid state has the potential to begin revealing the structural properties of the oligomeric precursors of fibrils [20], which appear to be the most medically relevant species [22]. Together, the combinations of solution and solid state NMR are powerful tools in the arsenal of techniques that are helping to improve our understanding of amyloid biology. We apologize in advance if for the sake of brevity we have omitted some studies in our review of this rapidly growing field. We would also be remiss if we failed to note the exciting strides magnetic resonance imaging has made towards the diagnosis of amyloid disease. As this subject is outside the scope of our review we refer the reader to some excellent articles [50-52].

2. Structure and Dynamics of Monomeric Amyloid Precursors

Since fibrils assemble from monomeric subunits the characterization of these initial states is critical for a complete understanding of amyloid biology. Key questions include: (1) are there conformational aspects of the monomers that make them intrinsically susceptible to aggregation? (2) Do factors such as sequence changes, solution conditions, or ligands shift the balance towards aggregated states? We will review NMR work on fibril precursors with a defined tertiary structure and precursors that are intrinsically unfolded. We will consider in detail NMR work on the precursors of the best-characterized amyloids Aβ and αS.

2.1. Folded Globular Precursors

NMR work on amyloidogenic proteins with folded native states has focused on identifying the structural and thermodynamic features that make these proteins susceptible to aggregation. NMR was the first technique to provide detailed structural information on the native state of prion protein [53]. The NMR structure showed that residues 126-230 of native PrP^c adopt a three-helix bundle structure comprised of some 44 residues, with only a short 8-residue segment of antiparallel β-sheet. The N-terminal part of the protein comprised of residues 23-125, which is important in pathology, was shown to be disordered [12,53]. The protein must therefore undergo extensive conformational rearrangements from a mostly α-helical PrP^C structure to the PrP^{Sc} fibrillar β-sheet structure recently demonstrated by EPR [54].

2.1.1. Hydrogen Exchange of Fibril Precursors with Folded Structures

Hydrogen exchange (HX) provides a sensitive probe of structural stability and slow-timescale dynamics [55,56]. Briefly, exchange-labile amide protons are protected by hydrogen-bonded secondary structure. The magnitude of protection depends on the stability of the structure to localized breathing motions and unfolding reactions that facilitate exchange [55]. Segments of structure with weak protection may correspond to the regions susceptible to conformational transitions that expose epitopes involved in aggregation. HX experiments have been used to characterize the native state stability landscape of transthyretin [57], lysozyme [58], and the hamster prion protein fragment 90-231 [59]. For transthyretin a correlation was noted between increased HX rates and amyloidogenicity, in mutants that either stabilize or destabilize the native structure [57]. Similarly, HX studies of amyloidogenic mutants of human lysozyme showed evidence for partial unfolding of the protein's β-domain, which is thought to provide an early scaffold for aggregation [58]. Studies of the prion protein were consistent with a stable core structure (residues 125-228) and a flexible N-terminus (residues 90-119). The first α-helix near the N-terminus showed lower protection to solvent exchange compared to the other helices [59]. The N-terminus of the molecule appears to be predisposed to adopt multiple conformations that may feature in the transition from PrP^C to the toxic PrP^{Sc} form [59].

2.1.2. Fast Timescale Dynamics of Folded Amyloid Precursors from NMR Relaxation Data

While hydrogen exchange experiments provide information on dynamics on slow timescales from seconds to years, ^{15}N NMR relaxation experiments are sensitive to flexibility on faster nano- to millisecond timescales. With prion protein, ^{15}N relaxation experiments provided evidence for increased flexibility of β-strand S1 and a partially folded hydrophobic cluster comprised of residues 113-128 at the N-terminus of PrP^C. Dynamics in this region may facilitate the conformational transition from PrP^C to the infective PrP^{Sc} form [59]. With another protein superoxide dismutase, ^{15}N relaxation data pointed to an increase in the flexibility of the β-barrel structure of a G93A mutant linked to familial amyotrophic lateral sclerosis. The transition from a closed β-barrel to a more open β-sheet structure in the mutant may promote fibrillogenesis by allowing intermolecular mispairing of β-sheets [60].

Flexibility on micro to millisecond timescales can be detected from relaxation dispersion experiments, or more commonly is inferred from NMR line broadening which is proportional to transverse (R2) relaxation rates [61]. Thus when a protein is highly unfolded, NMR resonances are sharp because of unhindered mobility. As nascent structure starts to form, this is often accompanied by slower motions that lead to larger transverse relaxation times and NMR signal broadening [62]. In some cases broadening is so severe that it results in complete loss of signals [63]. Care must be taken in these studies since weak intermolecular association [64] and fast hydrogen exchange [65] are two alternative mechanisms that can lead to similar resonance broadening. With appropriate controls in place, however, these types of experiments can be used to map regions of the unfolded state that are the first to fold into nascent structure. Such a study was used to characterize a partially folded intermediate of β2-microglobulin (β2m) formed at low pH [66]. The N- and C-terminal strands of β2m were found to be unfolded in this intermediate. Interestingly these same strands are weakly protected from hydrogen exchange in the β2m fibrils [67]. This suggests that the regions of β2m that are the first to fold in the intermediate also form the core of the fibril structure, supporting the hypothesis that fibril assembly occurs through partially folded states.

2.2. Intrinsically Denatured Precursors

While some fibrils are formed from globular proteins others start from intrinsically unfolded precursors. These proteins have properties typical of denatured proteins including random-coil like circular dichroism spectra, a lack of hydrogen exchange protection, extended hydrodynamic radii, and limited NMR chemical shift dispersion (Figures 2 and 3).

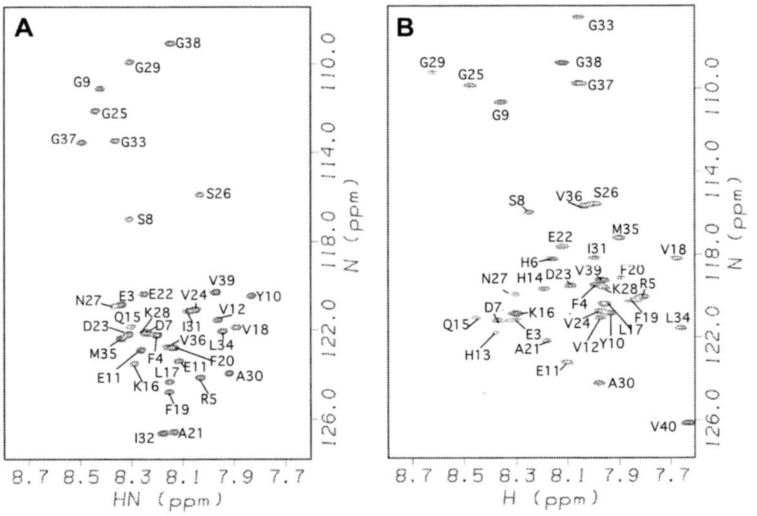

Figure 2. ^{1}H-^{15}N HSQC spectra of Aβ(1-40) (A) as a free monomer and (B) in the presence of 100 mM SDS micelles. Spectra were obtained at pH 7.2. The sample temperatures were 5 °C for free Aβ peptide, and 25 °C for micelle-bound Aβ. NMR assignments are from the literature [15,74,75,110] and our unpublished observations. The crosspeak for Val40 is outside of the region of the spectrum shown in (A).

Much of the NMR work on intrinsically unfolded precursors has sought to identify residual structure. Typically, a first step in the analysis of these proteins is to calculate differences in chemical shifts from random coil model compounds, as these differences are sensitive to residual α-helix or β-sheet structure (Figure 4). For Aβ, αS (see below), and IAPP [16] chemical shift differences are consistent with populations of α-helical conformers rapidly inter-converting with unfolded structure. There is considerable precedent for residual α-helix structure populated at low levels in the unfolded states of non-amyloidogenic proteins. This

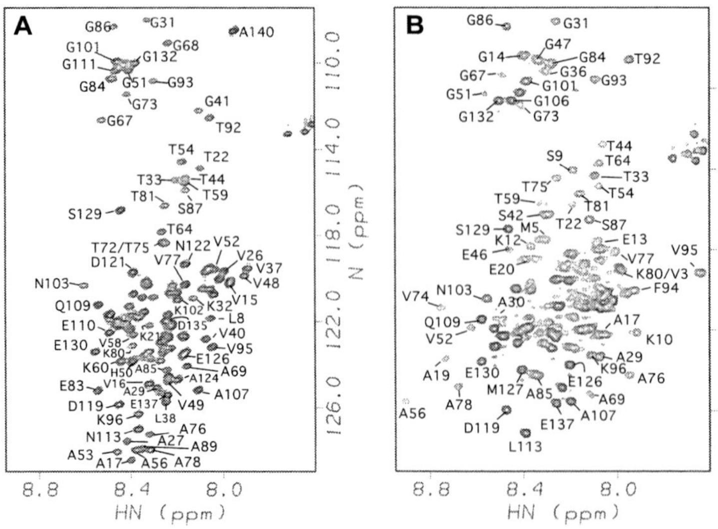

Figure 3. ^1H-^{15}N HSQC spectra of (A) free monomeric αS and (B) αS in the presence of 40 mM SDS micelles. Both samples were in 20 mM sodium phosphate buffer and the sample temperature was 35 °C. The pH was set to 6.4 to avoid signal losses in the free protein due to fast HX. NMR assignments are from the literature [14,106] and our unpublished observations. The crosspeak for Ala140 is aliased in (A) and not shown in (B).

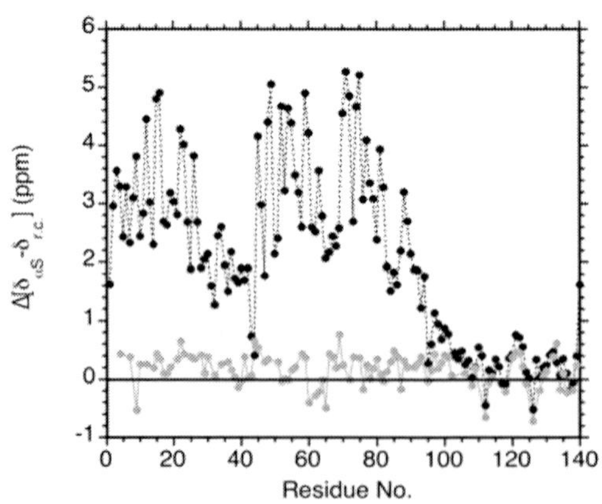

Figure 4. Differences in Cα chemical shifts from random coil [137] values. The solid line with gray points represents data for monomeric αS at pH 6.4 and 35 °C. The dotted line with black points denotes data for micelle bound αS (50 mM SDS micelles, 25 °C, pH 7.4) and is taken from the literature [106]. Positive chemical shift differences in this plot are indicative of α-helical structure. If we assume the first 100 residues in the N-terminal domain are 100% helical, we can estimate based on the size of the chemical shift differences that α-helix conformations are populated to levels of ~10% in the corresponding region of the monomer.

structure usually matches the locations of α-helices in the native structure [68]. The residual structure detected for amyloidogenic proteins with intrinsically unfolded precursors may reflect that the functional native states of these proteins are membrane-bound α-helical structures. Thus the location of residual α-helix structure in αS [14] and to a lesser extent Aβ [69] is similar to the location of stable α-helical structure when the proteins are bound to SDS micelles. The intrinsically unfolded protein tau on the other hand, shows segments with chemical shift deviations characteristic of extended-β conformations rather than α-helix. The preference for β-conformations is seen at sites involved in paired helical formation [70,71], and the population of β conformers increases on addition of polyanions or microtubules [71].

2.3. Monomeric States of Aβ

Structural studies of Aβ(1-40) and Aβ(1-42) in pure aqueous solution showed that the peptides were mostly unfolded, although three regions 8-12, 16-24 and 30-34 had conformational preferences sufficiently stable to calculate structure models. The segments adopted turn conformations. The longest segment with a cooperative structure, 16-24 consisted of a series of turns terminating in a helical turn between residues 20-24. The studies failed to identify structural differences between Aβ(1-40) and the Aβ(1-42) peptide [69], which causes much faster fibrillization. Similarly, oxidation of the Met35, which inhibits fibrillization has no effect on the structure of the Aβ monomer [15,72].

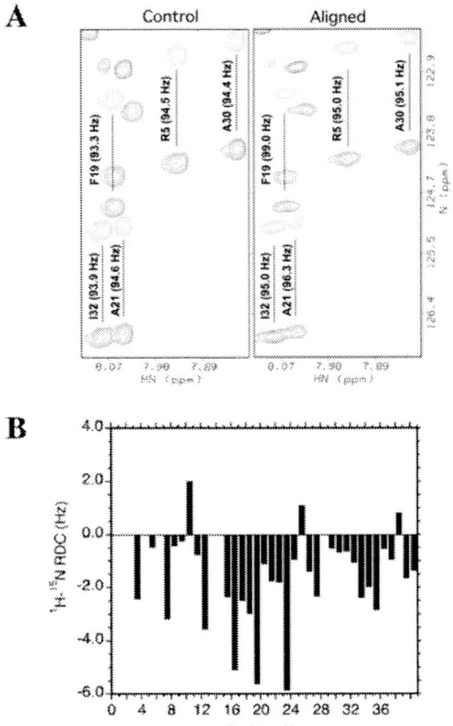

Figure 5. RDC data for Aβ(1-40) aligned in 19 mg/ml pf1 phage particles. Samples were at pH 7.3 and a temperature of 9 °C. (A) One-bond $^1J_{HN}$ splittings obtained in control (isotropic) and aligned (anisotropic) samples. The vertical lines labeled with the residue numbers connect the bottom right (black) and top right (gray) components of one-bond $^1J_{HN}$ TROSY quartets for each amide proton. (B) RDC values were obtained from $-1*(^1J_{aligned} - {^1J_{control}})$.

Residual dipolar couplings (RDCs) provide information on bond angles and the global anisotropy of a molecule. ^{1}H-^{15}N RDCs are all negative for Aβ(1-40) (Figure 5), typical of a mostly unfolded polypeptide [73]. ^{15}N relaxation data as well, indicate that monomeric Aβ(1-40) peptide has a high degree of flexibility characteristic of unfolded proteins [69,74,75]. A comparative study of ^{15}N relaxation rates in Aβ(1-40) and Aβ(1-42) found that the longer peptide exhibits reduced mobility at its C-terminus, which might play a role in its increased amyloidogenicity [69,74,75].

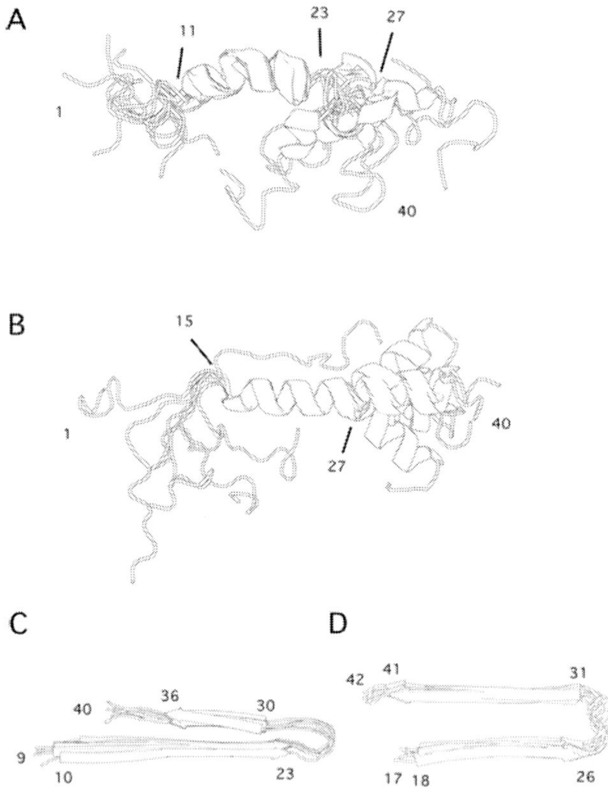

Figure 6. NMR structures of Aβ in various states. Each panel shows the first seven models of the respective NMR ensemble. (A) Monomeric Aβ(1-42) in 30% hexafluoroisopropanol / 70% H_2O. PDB code: 1Z0Q [138]. (B) Aβ(1-40) bound to SDS micelles. PDB code: 1BA4 [112]. (C) Aβ(1-40) fibril subunit [48]. The first 8 residues are disordered and were not included in the structure calculations. (D) Aβ(1-42) fibril subunit. PDB code: 2BEG [47]. The first 16 residues in the Aβ(1-42) fibrils were disordered based on qHX [47], and were not included in the structure calculations.

The most detailed structural model for monomeric Aβ was obtained for the Aβ(1-42) peptide in 30% hexafluoroisopropanol (HFIP)/ 70% water (Figure 6A). Although this structure was done in the presence of the structure stabilizing and possibly a structure-inducing alcohol HFIP, the structure has some interesting features that bear resemblance to structures of other states of Aβ (Figure 6B-D). The first ~10 residues of the peptide are disordered. This is followed by a stretch of α-helix structure running between residues 11 and 22. A disordered bend in the chain follows for residues 23-27. A second C-terminal helix runs from about residues 28-38. As the proportion of water increases and HIFP decreases, the

C-terminal helix becomes disordered while the N-terminal helix survives even at high water content.

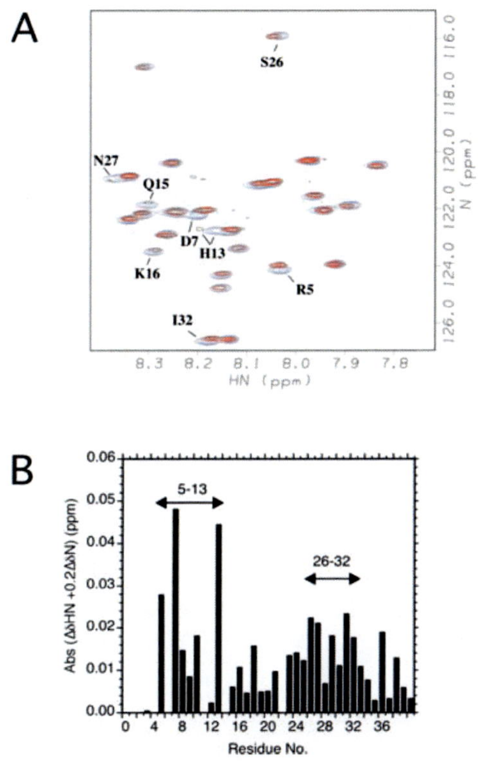

Figure 7. Binding of a low MW heparin fraction (MW 4-6 kDa, Sigma Catalog H8537) to Aβ(1-40). (A) Blue contours show the free Aβ peptide (0.12 mM in 20 mM sodium phosphate, pH 7.2, 9 °C). The red contours show the peptide in the presence of 2 mM heparin. Crosspeaks perturbed by the addition of heparin are labeled according to residue position in the sequence. (B) Absolute chemical shift differences between the free and heparin-bound Aβ(1-40) peptide. The most prominent feature is at the N-terminus of the molecule (residues 5-15), which contains all but one of the positive charges in the Aβ(1-40) sequence. A second feature occurs in the 26-32 region, near the salt-bridge stabilized hairpin turn that acts as a nucleus for fibrillogenesis [80].

The exquisite sensitivity of NMR to changes in magnetic environment makes it an excellent tool for monitoring ligand binding. Chemical shift perturbation studies have been used to map the location of a number of binding sites on Aβ peptides. Divalent Cu^{2+} and Zn^{2+} were proposed to be coordinated by the three histidines at the N-terminus of Aβ(1-40), and binding was established to occur with a K_d of about 5 μM [76,77]. The binding of the polyanion heparin was localized to the N-terminus of Aβ(1-40) which is important for neurotoxicity, and a K_d of ~1 mM was estimated per heparin disaccharide unit [78]. We observed similar results in our lab for the binding of a low 4-6 kDa MW heparin fraction to Aβ(1-40), except that we noted that in addition to the effects at the N-terminus, broadening and shift perturbations occur in a second region of the peptide centered on residues 26-32 (Figure 7). The affected residues were D23, S26, N27, G29, I32, and G33. A possible significance of these perturbations is that they are close to the hairpin turn found in the solid state NMR (ssNMR) structures of Aβ fibrils [79]. Moreover, studies have shown that the V24-28 turn is present in monomeric Aβ(1-40) where it may nucleate β-hairpin formation for subsequent assembly into fibrils [80]. If heparin stabilizes this nucleating turn structure it may explain why fibril formation rates are enhanced in the presence of GAGs [81]. Further structural and kinetic data will be needed to verify this hypothesis.

Recent experiments have characterized the interactions of Aβ peptides with each other, and with other proteins. Kinetic NMR experiments that followed the disappearance of Aβ(1-42) resonances due to aggregation were used to show that Aβ(1-40) monomers bind to Aβ(1-42) aggregates with a higher affinity than to Aβ(1-40) monomers [82]. The Aβ(1-40) monomers were suggested to confer protection against the aggregation of the more toxic Aβ(1-42) peptide [82]. A second protein αB-crystallin was found to compete with Aβ(1-40) monomers for aggregates, and to enhance the toxicity of Aβ(1-40) [83].

2.4. The Monomeric State of α-Synuclein

The monomeric state of αS has been extensively studied by NMR and has been shown to be largely unfolded. The N-terminal 100 residues of the protein consist of an amphipathic domain characterized by 7 imperfect repeats of the sequence KTKEGV. The hydrophobic NAC segment (residues 61-95), which is the most amyloidogenic part of the protein, is also located in the N-terminal domain. The C-terminal domain comprised of the last 40 residues has a high proportion of acidic residues [35]. NMR chemical shift analyses have shown low populations of α-helix structure in the N-terminal domain (Figure 4), while the C-terminal domain shows no conformational preferences [14]. Estimates of the fractional population of α-helix in the N-terminal domain are on the order of 10% at a temperature of 10 °C [14]. We obtain a similar estimate of α-helix population at 37 °C (Figure 4) showing that in contrast to a previous report [84] the α-helix content of αS stays about the same between 10 and 35 °C. ^{15}N relaxation data for monomeric αS are also typical of unfolded proteins devoid of long-range structure [85,86].

RDCs for αS are all negative, consistent with a predominately unfolded structure [85]. Deviations of RDC sequence profiles from those predicted for random coil models, have suggested that some residual long-range structure may be retained, involving interactions between the N- and C-termini of the protein [87,88]. This result, however, appears to not hold for different conditions and alignment media [89]. Additional evidence supporting the presence of a global structure stabilized by long-range interactions comes from paramagnetic relaxation enhancement (PRE) experiments, made possible by the incorporation of spin-labels at engineered cysteine positions [85,90]. In these experiments the unpaired electron of the spin label moiety causes NMR line broadening of protons that come within ~25 Å of the paramagnetic probe, thus affording a sensitive distance probe. It is worth noting, however, that because of the $\langle r^{-6} \rangle$ averaging properties of the PRE, the shortest contact distances (r) will dominate [91]. The PRE effects may therefore not reflect the true average structure but transiently collapsed conformations that could correspond to only a small fraction of the total. The PRE effects observed for αS have been interpreted in terms of a globular model, that brings together the N- and C-terminal domains of the protein through long-range interactions

[85,90]. This closed structure may protect αS from aggregation [85,90]. A number of different positively charged polyamines, such as spermine and putrescine bind to the acidic C-terminal domain [92]. Presumably these polycations enhance the rates of αS aggregation by disrupting the 'auto-inhibitory' interaction between the N- and C-terminal domains [85,92]. The resultant opening of the structure may explain how polycations enhance the rate of αS aggregation. In addition to polycations, high temperature also appears to disrupt long-range interactions leading to a fully unfolded conformation [85].

Under the pH 7.4 conditions typically used for NMR studies of αS, amide proton signals from the first 100 residues in the N-terminal domain disappear when the sample temperature is raised from 10 °C to 35 °C. The amide protons from the C-terminal domain are still observed [84]. Line broadening of NMR resonances from the same region of αS occurs when the intrinsically denatured protein folds into an α-helical structure in the presence of micelles [14]. This led to the conclusion that α-helical conformations in αS are promoted with increasing temperature [84]. In contrast to the results *in vitro*, nearly the full complement of ^{1}H-^{15}N correlations were seen for αS encapsulated within living *E. coli* cells at 37 °C [93]. This observation was ascribed to an unfolding of the presupposed α-helical structure, by the "molecular crowding" that occurs in living *E. coli* cells [93]. These conclusions have important ramifications for our understanding of αS. (1) Preformed α-helical structure in αS monomers would facilitate association of the protein with membranes, and would hinder the transition to aggregative β-sheet structure. (2) If the cellular environment of *E. coli* had a vital influence on modulating the conformation of αS, it would cast serious doubt on the relevance of structural studies performed on the purified protein in solution.

Our results (Figures 8 and 9) are inconsistent with the conclusions that (1) αS undergoes a significant increase in α-helical structure with increasing temperature and that (2) molecular crowding alters the conformation of αS [65]. We found at least three lines of evidence that indicate hydrogen exchange rather than a conformational transition leads to the temperature-induced peak losses in ^{1}H-^{15}N HSQC spectra of αS at pH 7.4 [65]. (1) Although correlations are lost in ^{1}H-^{15}N HSQC spectra at high temperature, ^{1}H-^{13}C HSQC spectra are unchanged between 10 and 35 °C (Figure 8). Unlike amide protons, carbon-bound protons are not labile to hydrogen exchange. This result is in marked contrast to the loss of NMR signals from both ^{15}N- and ^{13}C-attached protons when αS folds into α-helical structure in its large molecular weight complexes with lipid micelles [65]. (2) As noted by others [94], the steep pH dependence of line broadening in ^{1}H-^{15}N HSQC spectra is consistent with the $10^{\Delta pH}$ dependence of hydrogen exchange rates. The entire spectrum of αS is recovered at 35 °C when the pH is lowered from 7.4 to 6.4 (Figure 3A). Conversely, most resonances disappear at the lower temperature of 10 °C, when the pH is increased from 7.4 to 8.4 [65]. (3) We were able to measure hydrogen exchange rates for all sites in the protein at 15 °C using CLEANEX NMR experiments [95]. The observation that rapid exchange affects all amide protons at 15 °C is consistent with a loss of amide protons at the higher temperature of 35 °C due to HX, since exchange rates are expected to move from the slow to the fast regime with increasing temperature. The HX rates we determined at 15 °C [65] provide an explanation for the selective disappearance of amide proton signals from the N-terminal domain. Exchange rates from the C-terminal domain are slower as a result of the high density of negative charges in the amino acid sequence of this region. CLEANEX experiments [95] at high salt further confirm that this is an electrostatic effect [65]. Finally, based on Cα chemical shift data we can conclude that between 10 and 35 °C the amount of α-helix structure in monomeric αS is maintained between ~5 and 10 % (Figure 4).

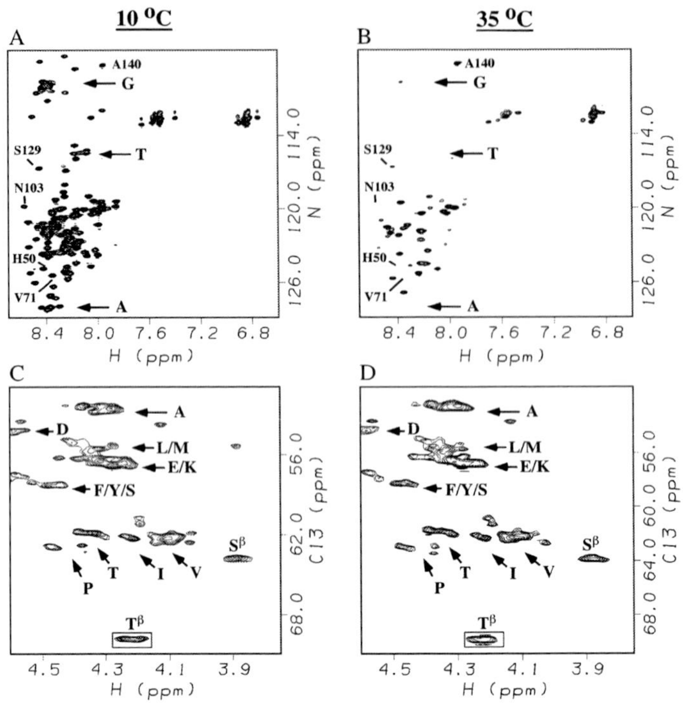

Figure 8. NMR spectra of αS at 10 °C (A,C) and 35 °C (B,D). All spectra were for samples at pH 7.4. The panels show ^{1}H-^{15}N HSQC spectra (A, B) and portions of Hα-Cα selective ct-^{1}H-^{13}C HSQC spectra (C,D). Whereas ^{1}H-^{15}N correlations are lost with increasing temperature due to fast HX (compare A and B), ^{1}H-^{13}C correlations are nearly unchanged over the same temperature range (compare A and B). This demonstrates that the loss of amide proton NMR signals with increasing temperature at pH 7.4 is due to fast HX, rather than a purported conformational transition to α-helical structure [84,93]. Selected residue-type and sequence-specific assignments are denoted with single-letter amino acid codes. Labels without superscripts in C and D indicate Hα-Cα connectivities. Boxed correlations are those with negative phases in the ct-^{1}H-^{13}C HSQC spectra.

Since the disappearance of ^{1}H-^{15}N HSQC signals at high temperature and pH 7.4 *in vitro* can be attributed to hydrogen exchange rather than a conformational transition, there is no need to invoke cellular factors to explain the unfolded conformation of αS when it is encapsulated within living *E. coli* cells. The most plausible explanation for the persistence of ^{1}H-^{15}N HSQC signals from αS in living *E. coli,* is that in contrast to the *in vitro* studies at pH 7.4, the protein is likely to experience a more acidic pH near or below pH 7.1 in the intracellular medium [96]. A drop in pH is typical of *E. coli* cultures grown to high-density, such as those used for *in vivo* NMR experiments. Lowed pH results in slower hydrogen exchange rates [56], so that a higher proportion of backbone amide proton signals survive in the NMR spectrum. *In vitro* all of the missing signals at 35 °C could be recovered by lowering the solution pH over the narrow range from 7.4 to 6.4. The result that the purported molecular crowding agent bovine serum albumin (BSA) results in a recovery of amide proton signals from αS is also likely to be due to a drop in pH, when a 300 mg/ml concentration of the acidic BSA protein is added to an αS solution. In our hands adjusting the pH of the mixture back to 7.4 after the addition of BSA, results in ^1H-^{15}N HSQC spectra comparable to those of free αS, where amide proton signals from the N-terminal domain are lost due to fast hydrogen exchange (Figure 9). These results further emphasize that the initial monomeric state of αS is highly unfolded, with a population of at most 10% α-helical conformers [65].

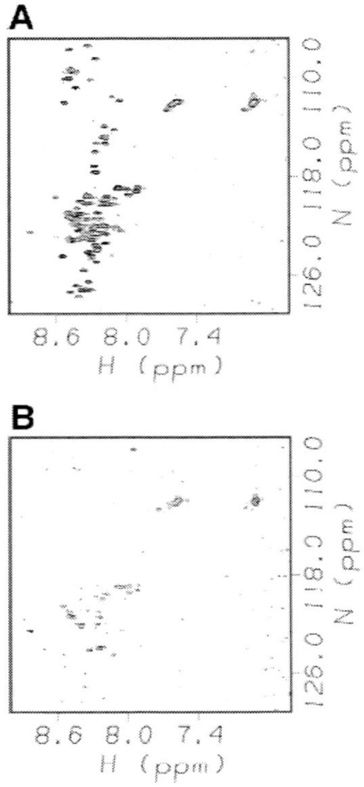

Figure 9. ^1H-^{15}N HSQC demonstrating that the effects of high concentrations of bovine serum albumin (BSA) on NMR spectra of αS are due to changes in solution pH rather than 'molecular crowding' [93]. Spectra were obtained at 35 °C on a sample containing 0.25 μM αS in a 20 mM sodium phosphate buffer that had an initial pH of 7.4. (A) When a large excess of BSA was added (320 mg/ml; ACROS catalog number 61191-0010), the acidic protein overwhelmed the buffering capacity of 20 mM phosphate and the solution pH dropped to 5.7. Most of the amide protons expected from the amino acid sequence of αS were seen under these conditions because of slow HX at this acidic pH [56]. (B) When the same sample was readjusted to pH 7.4 by addition of NaOH, however, ^1H-^{15}N correlations were lost due to faster HX at the more basic pH [56].

The availability of NMR assignments for monomeric αS [14] has made it possible to map the binding sites for a variety of ligands. Amongst the transition metals characterized, αS has the highest affinity for Cu^{2+} with a K_d of ~0.1 μM [97]. The high affinity Cu^{2+} site is located in the N-terminal domain, and residue His50 forms part of the binding site [97,98]. Lower affinity Cu^{2+} sites are formed by the acidic residues from the C-terminal domain [99], primarily the Asp119-Ala124 segment, which also binds other metals such as Fe^{2+}, Co^{2+}, Mn^{2+} and Ni^{2+} with low affinity, (K_d ~1 mM). The C-terminal acidic domain of αS binds polyamines including spermine, putrescine, spermidine, and BE-4-4-4-4, with weak affinity

(K_d values in the 0.4 to 10 mM range) [92]. The binding affinity for polycations correlates with their efficacy in promoting αS aggregation [92]. An NMR chemical shift perturbation study suggested that αS bound to Aβ in the presence but not in the absence of SDS micelles [100]. It now appears, however, that some of the changes in chemical shifts on which this conclusion was based were an artifact of different pH values for the stock solutions of αS and Aβ used to prepare the mixture [101]. This together with the BSA results described earlier, highlights the importance of accurate control of experimental variables in titration experiments with intrinsically unfolded proteins, since small variations in solution conditions often result in large changes in NMR spectra.

In addition to the wild type αS, NMR has also been used to study mutations that cause early-onset Parkinson's disease, and the non-amyloidogenic homologues βS and γS. Based on NMR chemical shift data, the A30P mutation disrupts residual structure at the N-terminus of αS because of the inability of the substituted proline to participate in an α-helix [86]. A second A53T mutant was shown to have little effect on either chemical shifts or ^{15}N relaxation parameters of the unfolded monomeric state [86]. Similarly, the E46K mutation has subtle effects, if any, on the monomer [102].

Recently NMR studies were extended to the β and γ homologues of αS. The homologues aggregate less readily, and are not linked to neurodegenerative disease [89,94,103]. Chemical shift data suggest that γS has a lower propensity for α-helix structure compared to αS [89,103], while for βS there appears to be disagreement about the interpretation of the small chemical shift differences from random coil values [89,94]. PRE data suggest that βS and γS do not form the transiently collapsed globular conformations seen with αS [89,103].

3. Modeling Membrane Bound Forms of Amyloidogenic Proteins

3.1. Structural Investigations of Amyloidogenic Proteins in Their Micelle Bound States

A number of amyloidogenic proteins such as αS may exert their normal function through membrane bound states [34]. Moreover, the toxicity of amyloidogenic proteins may be due to the effects of oligomers on membranes [22]. It is therefore important to characterize the membrane bound states of amyloidogenic proteins. Unfortunately, complexes formed between proteins and lipid bilayers are sufficiently large and complex to be unapproachable by most solution NMR techniques [42,104]. Consequently, solution NMR studies have employed membrane mimetics such as micelles. Even with these model systems some lipids make micelles that are too large for NMR. The vast majority of investigations have used sodium dodecyl sulfate (SDS) micelles. There are some caveats worth mentioning about these systems. (1) Micelles have a lipid monolayer structure. Bilayers such as those of unilamellar vesicles offer a more realistic representation of a membrane. Unfortunately unilamellar vesicles are too large for NMR. (2) SDS is a strong denaturant for most folded proteins (part of the reason for its use in SDS-PAGE). Unfolded proteins with amphipathic sequences, however, readily adopt α-helical structures on the surfaces of SDS micelles [105]. This raises the question of whether the structure observed with SDS micelles represents the membrane-bound state of the proteins or non-native structures induced by the detergent. This question can be resolved through careful controls such as those done for αS, where the protein was studied both with micelles and unilamellar vesicles to see if the choice of membrane mimetic affects structure. Studies with micelles formed from different types of lipids can also be used to determine to what extent the nature of the lipid affects protein structure [104,106]. (3) Another concern is that the size of the micelle could affect the structures observed. Thus the NMR structure of αS when bound to SDS micelles (Figure 10) consists of two long curved α-helices separated by a flexible hinge between residues 38 and 44 [104,106]. Both the break in the α-helices and their curvature may be an artifact of the inability to accommodate an uninterrupted α-helix on the surface of an SDS micelle [107].

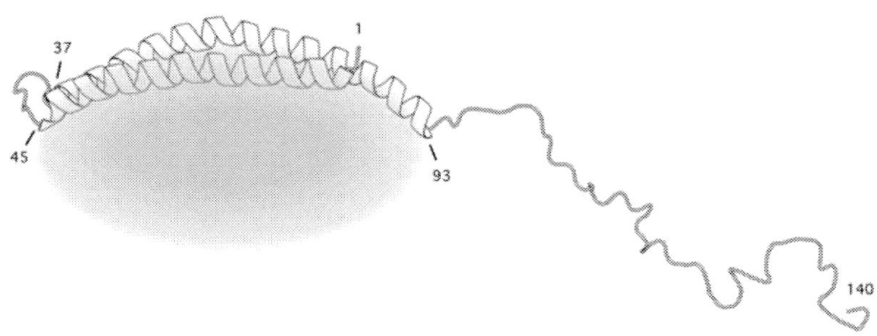

Figure 10. NMR structure of αS bound to SDS micelles. PDB code: 1XQ8 [104,116]. The gray ellipsoid is a representation of an SDS micelle. The C-terminal domain comprised of residues ~100-140 is unstructured in the micelle bound state.

In spite of the reservations above, SDS micelles have led to useful insights, and to the only structural models of the membrane-bound forms of amyloidogenic proteins. Examples include a SDS micelle-bound structure of calcitonin that shows the 32-residue hormone adopts a flexible amphipathic α-helix from residues 13-21 [108]. A K19 fragment containing three repeats from the C-terminal domain of the microtubule associated protein tau was recently characterized at the level of delineating the secondary structure of the SDS micelle bound protein [109]. Three stretches of α-helix were identified based on chemical shit data, $^3J_{HNHA}$ coupling constants and HN-HN NOEs. The location of α-helical structure in the micelle bound state matched the location of residual α-helix structure in the monomeric state. PRE data were used to show that the tau construct lies along the surface of the SDS micelle, without traversing the interior [109]. To rule out effects specific to SDS micelles, the fragment was shown to have similar structural properties when bound to a micelle formed from by lyso 1-palmitoyl phosphatidylglycerol (LPPG).

3.2. Micelle Bound Structures of Aβ Peptides

In the presence of SDS micelles Aβ(1-40) folds into two α-helices (Figure 6B), that lie along the surface of the micelle [110,111]. The α-helices comprised of residues 15-24 and 28-36 are separated by a kink or hinge between residues 25-27. The first ~14 residues and the last ~4 remain unstructured in the micelle bound form of the peptide [110-112]. Although the relative orientation of the two helices is very poorly defined, they are in proximity and are arranged in roughly an antiparallel hairpin [111,112]. The more amyloidogenic Aβ(1-42) peptide shows no major structural differences from Aβ(1-40) in its micelle bound form [111]. The SDS micelle-bound Aβ(1-40) structure is pH sensitive, and is more unfolded under slightly basic (pH 7.9) than acidic (pH 5.1) conditions [112]. In addition to SDS, Aβ has been studied with other micelle systems. Mixed SDS/ganglioside micelles [113] and GM1 micelles [78] were used to map the Aβ binding site for sialic acid carbohydrate moieties in the GM1 ganglioside headgroups. The GM1 lipids were studied because they have an inhibitory effect on amyloid formation [78].

An interesting feature of the micelle-bound structure of Aβ is that the topology is similar to that of Aβ in amyloid fibrils [47], and to the residual structure observed within the unfolded Aβ monomers [69]. In Aβ(1-42) fibrils, the monomer subunits form an anti-parallel β-hairpin consisting of segments 18-26 and 31-41, and the first ~15 residues are disordered [47,114]. In spite of the transition from α-helix secondary structure in the micelle-bound state to a β-sheet structure in the amyloid fibrils, the hairpin between adjacent elements of secondary structure, and the disorder of the first ~15 residues at the N-terminus are similar in the micelle-bound structure. This suggests the turn centered around residues 25-27 [80] might determine the topology of both the micelle-bound and fibrillar Aβ structures. Turns may be as important to fibril structures as they are to the native states of proteins [115].

3.3. Micelle-Bound αS

When bound to SDS micelles, the N-terminal amphipathic domain of αS adopts an α-helical structure while the acidic C-terminal domain remains unstructured (Figure 10) [14,104,106,116]. Biochemical and NMR studies with micelles formed from a variety of lipids [104] and with small unilamellar vesicles [106] were used to verify that the α-helix structure observed with SDS micelles is a general feature of membrane-bound αS.

A high-resolution structure of SDS micelle-bound αS was obtained by NMR [104]. The protein folds into two long curved helices, interrupted by an ordered extended linker between residues 38-44 (Figure 10). The hairpin topology of the α-helices in the micelle-bound αS structure is well defined by RDC and PRE data (Figure 10). NOEs between the α-helices, however, were not observed [104]. The closest distance between the two helices in the resulting NMR structure is about 9 Å. The rotameric states of side-chains from the two α-helices were found to be averaged [104], also suggesting that there is no packing of side-chains between the two α-helices. The hairpin orientation appears to be primarily determined by the conformational properties of the linker between the α-helices.

Alternatively, the anti-parallel arrangement of the α-helices may be a consequence of the sizes of SDS micelle, since a single unbroken α-helix would have a larger diameter than the micelle [104]. Presynaptic vesicles are ten times larger than SDS micelles, and EPR evidence points to an uninterrupted α-helix structure for αS on larger membrane mimetics such as unilamellar vesicles [107]. The bend or curvature of the individual α-helices (Figure 10) may also be a consequence of the accommodation of the structure to the curvature of the SDS micelles [104]. Nevertheless, the perturbation induced by the limited size of the SDS micelle is relatively minor, and limited to the conformation of the hinge region between residues 38-44. The portion of the protein that is α-helical is likely to accurately reflect the membrane bound structure [104,106].

The α-helices lie along the surface of the SDS micelle [104]. Experiments transferring saturation from the resolved resonances of SDS to the protein, were used to assess the orientations of the α-helices with respect to the micelle surface, and the depth of lipid penetration for each residue [116]. A positive correlation was found between the degree of immersion at a given site, and reduced dynamics as measured from ^{15}N relaxation and/or hydrogen exchange [116]. Although the overall anti-parallel topology of the structure appears to be well-defined [104], ^{15}N relaxation parameters and RDCs suggest that the α-helices experience motion on a range of timescales (nanoseconds to seconds). Regions 30-37, 65-70, and 83-89 in particular show evidence for large amplitude dynamics on sub-nanosecond timescales.

The three disease-linked mutations A30P, E46K and A53T have also been characterized in their SDS micelle bound states [86,102,116]. The A30P mutation leads to a localized disruption of α-helix structure, an increase in dynamics near the site of the mutation [86,116], and an apparent lower affinity for lipid vesicles compared to the wild type [86,116]. Based on chemical shift data in the micelle bound state, the A53T mutation affects only residues directly adjacent in the sequence [86,116]. The E46K mutation affects residues some 15 positions on either side of the substitution, suggesting that it perturbs the α-helix structure [102].

There are many interesting facets of the micelle bound conformations yet to uncover, and our lab is actively working on some of these issues. What forces stabilize the micelle bound structure? How does the structure respond to changes in conditions? How is the protein structure affected by changes in lipid composition or by the properties of the micelles? Are there time-dependent changes in the micelle bound structure that could be related to the mechanisms of protein aggregation?

4. Amyloid Fibrils by Quenched Hydrogen Exchange

Amyloid fibrils and oligomeric intermediates are too large for direct solution NMR studies. Nevertheless, it is sometimes possible to transfer information from these large aggregates to monomers amenable for NMR studies. In rare cases structural parameters such as chemical shifts, NOEs, coupling constants, or RDCs may be transferred from aggregates to monomers through exchange, if the kinetics of dissociation are fast enough to retain magnetization before it is dissipated through relaxation processes [6,17,83,117]. Another approach is afforded by quenched hydrogen exchange experiments. These employ NMR of denatured monomeric states to read out the locations of amide protons trapped by hydrogen bonded structure in the fibrils, and can quantitatively determine the amount of protection against solvent exchange afforded by the stability of the fibril structure [44].

4.1. Quenched Hydrogen Exchange Methods for Studying Amyloid Fibrils

The use of dimethyl sulfoxide (DMSO) quenched hydrogen exchange (qHX) was pioneered by Roder's group to look at peptides that exchanged too fast to measure in water [118]. Our group was the first to publish an application of this method to amyloid fibrils formed by the model *E. coli* protein CspA [44]. The technique, which is outlined in Figure 11 takes advantage of two properties of DMSO. (1) DMSO is a strong denaturant and can dissociate some (but not all) types of amyloid fibrils. (2) DMSO is an aprotic solvent so that amide protons of unfolded proteins can survive sufficiently long in this solvent to be recorded in NMR experiments.

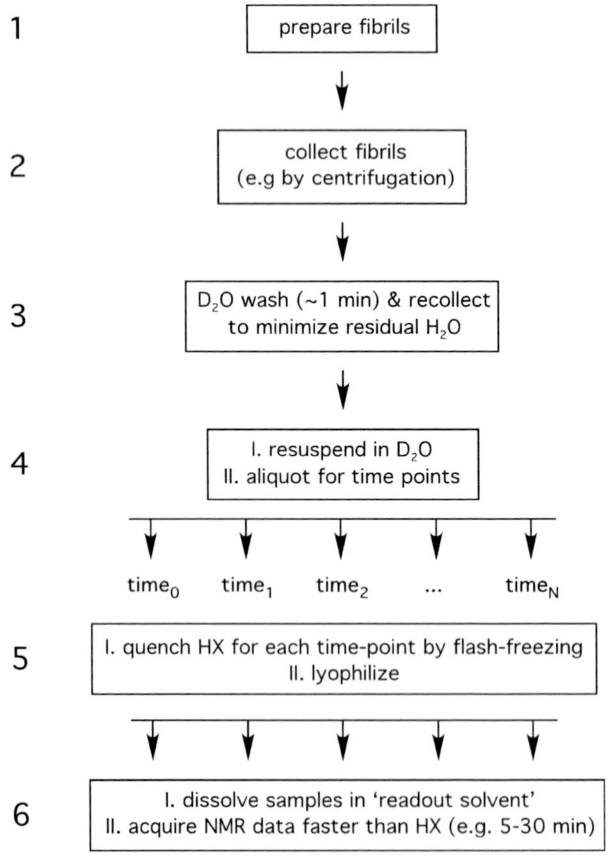

Figure 11. Scheme for qHX experiments of amyloid fibrils.

A number of the factors that can affect the success of qHX NMR experiments are worth considering. (1) Sufficient fibrils must be available for NMR experiments (step 1 in Figure 11) and there must be a way to collect the fibrils. Some fibrillization reactions are extremely slow. Even though fibrils may be detectable by EM, the reaction may not have gone to completion and most of the protein may persist in a soluble form. The progress of fibrillization should ideally be checked by multiple techniques [20,119] such as thioflavin T fluorescence assays, A280 of the supernatant, SDS-PAGE of soluble and insoluble fractions, and EM. Once sufficient fibrils are prepared they must be separated from the rest of the sample for D_2O exchange. Typically this is done by centrifugation [44,47]. For some fibrils that form a poorly sedimentable viscous gel, filtration through a MW cut-off filter (e.g. 100 KDa) may be a preferable alternative. (2) The protein must be amenable to NMR in the 'readout solvent' (Figure 11, step 6). Suitability is best tested on fresh protein samples that have never been fibrillized. An example of this kind of solvent screen for αS is shown in Figure 12. Note that the state the protein adopts in the readout buffer (e.g. helical, oligomeric) does not matter for the qHX experiments. The only requirement is that the amide protons are trapped in the readout solvent, and that their signals can be resolved. (3) Another criterion, is that the fibrils must be soluble in the readout solvent. This can be tested by comparing NMR spectra of fresh protein, and fibrils dissolved in the same readout solvent. For these experiments the fibril sample should be protonated (e.g. never exposed to D_2O) so that the intensities of amide proton signals are not perturbed by HX. This control can also be used to test if lyophillization affects the fibrils. (4) Back-exchange from residual H_2O in the fibril sample can artifactually appear as weak protection during the readout experiments. A simple way to minimize traces of H_2O is by washing the fibrils in D_2O prior to the exchange experiments (step 3 in Figure 11). (5) A final concern is that exchange in the readout buffer should be negligible during the NMR experiments used to detect amide proton intensity trapped by the fibril structure. This can be verified by determining the exchange rates of the protein in the readout solvent. For example, if NMR detection takes 5 minutes and the half-life for exchange is 3 hr, exchange during the detection period will have no effect. In cases where the time for NMR experiments becomes comparable to that for HX, an elegant analytical method has been developed to correct for this [47].

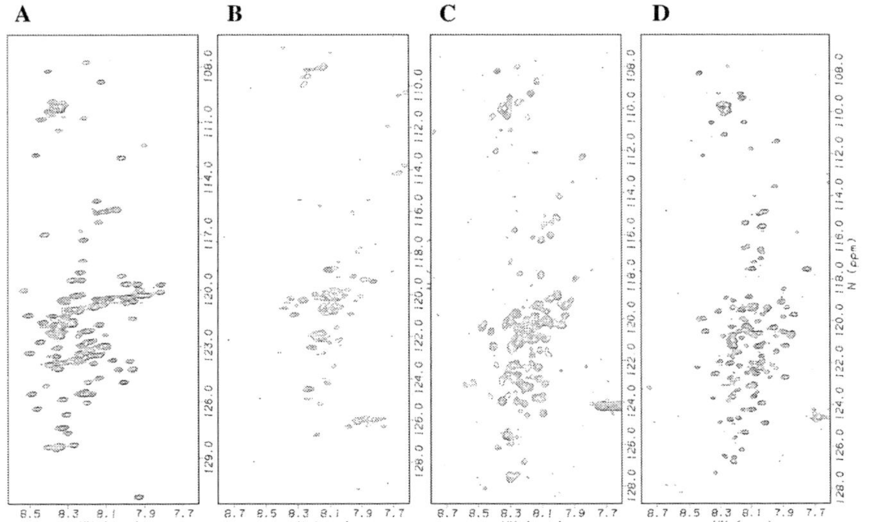

Figure 12. Tests for the applicability of different solvents to qHX experiments with αS. In each case freshly prepared, lyophilized, ^{15}N-labeled αS was dissolved in the indicated solvent. (A) Control conditions: water, 20 mM phosphate, pH 7.2. (B) 90% DMSO, 10% H_2O, pH 1.5. (C) 70% DMSO, 30% H2O, 6 M urea, pH 2.0 (D) 50% acetonitrile, 50% water, pH 1.5. All samples were at 25 °C.

Results from our original studies with the CspA model system are shown in Figure 13. Surprisingly, and in marked contrast to typical native proteins, all of the backbone amide protons were protected in the CspA fibrils and exchanged with close to the same rate. We interpreted this observation in terms of an exchange mechanism that involves dissociation or shedding of monomers from the fibrils [44]. The qHX technique has now been applied to fibrils formed from a number of medically important proteins and the results of these investigations are reviewed below. Some fibrils show a mixture of protected and unprotected segments. Typically, however, the protected cores of fibrils show remarkably uniform exchange rates and protection factors [44,47,67,120]. This in contrast to globular proteins where HX rates can vary significantly along the length of a single α-helix or β-strand. We believe that the relatively uniform HX rates observed for residues protected in fibrils, reflects the extensive intermolecular hydrogen bonding that stabilizes this type of structure. Localized breathing or unfolding motions that allow exchange in globular proteins are probably exceedingly rare in fibrils. In order for amide exchange to occur a very cooperative network of hydrogen bonds must be broken. This may only occur when fibrils dissociate or shed monomers, which then exchange at rates much faster than in the fibrils [44].

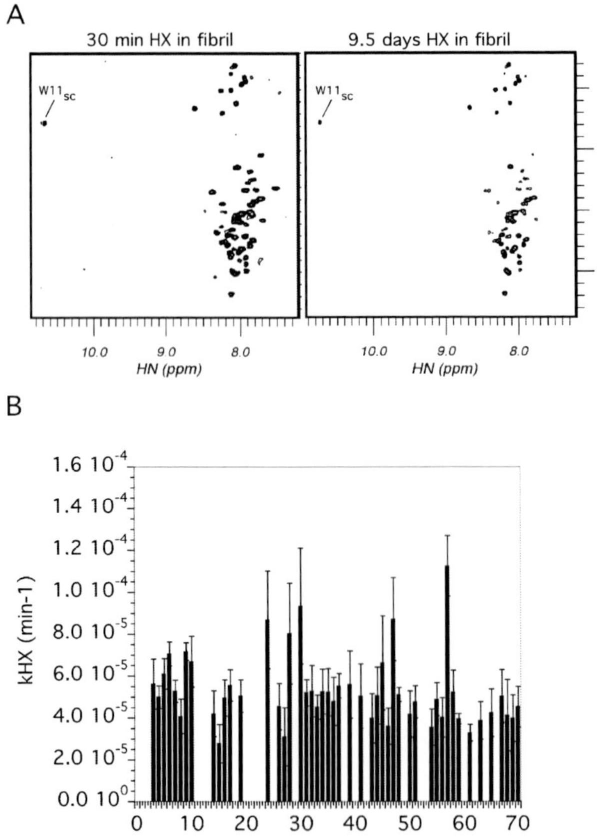

Figure 13. qHX of CspA fibrils [44]. (A) Time-points from the beginning and end of the fibril HX experiment. (B) Exchange rates for individual residues in the fibrils.

4.2. Quenched Hydrogen Exchange of Aβ Fibrils

Fibrils formed from the Aβ(1-40) peptide confer strong HX protection to residues 16-36 that constitute the hydrogen bonded core of the β-hairpin structure [121]. Residues G25 and S26 in a β-hairpin turn are only weakly protected. Similarly, the first ~15 and last 4 residues at the ends of the peptide exchange rapidly [121].

There are subtle but significant differences in the HX protection of the Aβ(1-40) and Aβ(1-42) fibrils [47,122,123]. In Aβ(1-42) fibrils, protection extends further to residue 40 at the C-terminus. By contrast, residues 4-14 at the N-terminus show weaker protection in Aβ(1-42) than in Aβ(1-40). Another difference occurs in the hairpin bend where relative protection appears to be weaker for residues 25-27 in Aβ(1-42) compared to Aβ(1-40) [122]. In the case of Aβ(1-42) fibrils, it was possible to determine exchange rates in the fibrils rather than proton occupancy at one or two time-points [47]. The reported rate constants for the slowest exchanging amide protons in the β-sheets of the Aβ(1-42) fibrils, correspond to half-lives of ~8 years [47]. The hydrogen exchange data for the Aβ(1-42) fibrils were used as hydrogen-bond restraints to calculate a structural model of the fibril structure [47]. It is worth mentioning that while protection data for fibrils have typically been interpreted in terms of structure [47,121,122], weak protection by itself need not imply a lack of structure. HX occurs through transiently disrupted structures that expose amide protons to the solvent. These transient conformations provide an avenue for exchange even if they constitute only a minute fraction of the conformational ensemble [55].

4.3. Quenched Hydrogen Exchange of Fibrils Formed from Other Amyloidogenic Proteins

In addition to Aβ, qHX has been used to characterize the fibrillar states of four proteins with folded native state structures: apolipoprotein II [124], β2-microglobulin [67,125], transthyretin [126] and cystatin [120]. Apolipoprotein II exists as an α-helical protein in its lipid bound form. Quenched hydrogen exchange studies of fibrils formed from this protein detected two protected cores formed by residues 19-37 and 57-74 [124]. These cores match the segments of the protein that are protected from proteolysis in the fibrils. For β2-microglobulin (β2m), a core of ~60 sites in the middle of the 100-residue protein were protected in the fibril, after 8 days in D_2O. By contrast, the N- and C-terminal ~15 residues were unprotected [67]. For β2m, the fibrils were formed and exchange was studied at pH 2.5 [67]. Because β2-microglobulin fibril morphology is very sensitive to solution conditions [67], the physiological relevance of the data obtained at pH 2.5 remains an open question. Additional studies were carried out for an 'immature form' of β2m fibrils obtained at pH 2.5 in the presence of high salt concentrations [125]. The immature fibrils represent a dead-end, off-pathway intermediate since they cannot seed the formation of the mature fibrils. Compared to the mature fibrils, the core protected in the immature fibrils was smaller spanning a region of about 30 residues in the middle of the protein [125]. The protection pattern for either type of fibril did not match that of the native state [67,125]. By contrast, protected regions in fibrils of transthyretin matched the locations of 6 of the 8 β-strands in the native structure, suggesting that the fibrils may form through an intermolecular rearrangement of native β-sheet pairing [126]. In cystatin B, the loop/strand protection pattern of the native protein was closely matched in the fibril state for β-strands 2 through 5 [120]. The first β-strand and subsequent α-helix were unprotected, suggesting a role for domain swapping in the fibril structure [120].

Besides NMR, mass spectrometry offers a complementary approach to investigate hydrogen exchange, and readers are directed to an excellent review on the applications of this method to investigations of fibrils [127]. We anticipate that the NMR applications for qHX experiments are only beginning to realize their full potential. Future applications include following the kinetics of fibrillization [125], the effects of sequence changes on exchange [122], the effects of various cofactors on exchange [128], and the positional dependence of hydrogen exchange (for example to investigate if protection is similar for monomers incorporated at the ends and in the middle of fibrils). It is further anticipated that experiments analogous to the native state hydrogen exchange technique for globular proteins [55,129] should make it possible to obtain quantitative information on the thermodynamic stabilities of fibrils, and the cooperativity of structure stabilization.

5. Amyloid Fibrils and Intermediates by ssNMR

Some of the most remarkable progress in the structural characterization of amyloid fibrils has come from ssNMR. The technique lacks the size limitations of solution NMR and the requirement for a crystalline three-dimensional lattice of X-ray crystallography. Isotopic labeling makes it possible to establish distance contacts through dipolar couplings, which together with torsional angle restraints from secondary Cα and Cβ chemical shifts, qHX data, and information from other techniques [47] can be combined as input in simulated annealing protocols to calculate NMR structures [130].

Prototypical examples of fibril structures solved by ssNMR include Aβ fibrils [47,48], a WW domain that is co-deposited with huntingtin in Huntington's disease [49], and a 22-residue fragment of β2 microglobulin [131]. Based on these structures, a general model of amyloid protofilament structures has been proposed [49]. The monomer adopts a β-hairpin structure (non-native in case of the WW domain), oriented perpendicular to the long fibril axis (Figure 14A). The structure is stabilized by a "steric zipper", which in this model is formed by interdigitating side-chains across the β-hairpin (Figure 14B, S=0). The hairpin lacks intramolecular β-sheet hydrogen-bonds but uses intermolecular hydrogen-bonds to pair adjacent monomer subunits stacked in a parallel β-sheet along the length of the fibril [49]. Recent ssNMR structural work on Aβ fibrils has provided at least one exception to the generality of this model. In contrast to the intramolecular side-chain packing of the WW domain fibril structure [49] the Aβ fibrils use intermolecular packing [47,48], which leads to a staggered arrangement of the β-strands (Figure 14B, S=2). As models of fibril structures become more numerous and detailed, the generic model of fibril structure will need to be refined and extended.

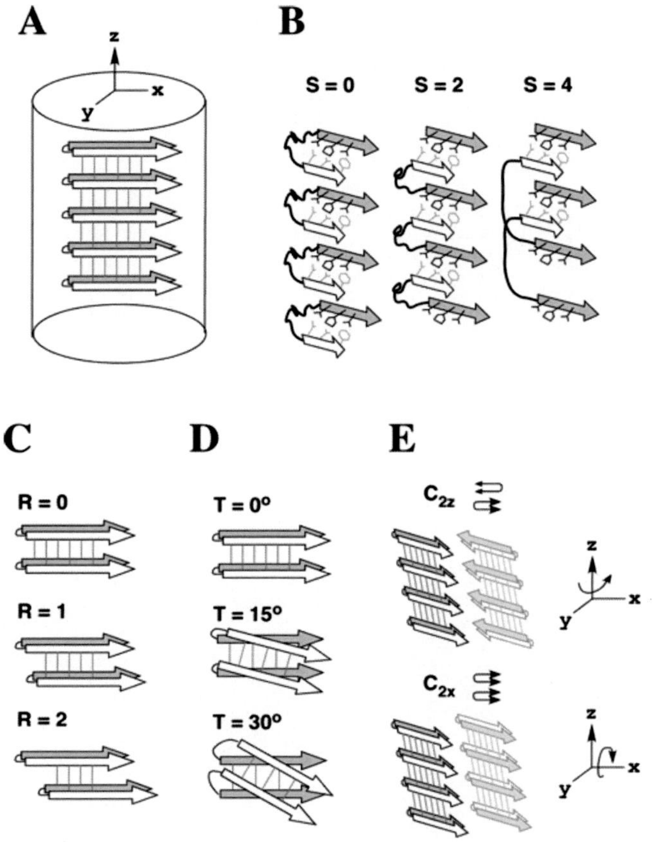

Figure 14. Schematics of amyloid fibril structures. In all diagrams the z-axis was arbitrarily chosen to run along the length of the fibril. The β-strands are thus perpendicular to the z-axis and the β-sheet hydrogen bonds parallel to it. (A) Generalized model of fibril structure. The monomer subunits fold into a β-hairpin but the backbone forms no intramolecular hydrogen bonds. Instead, intermolecular hydrogen bonds (thin vertical lines) are formed between monomers in an infinite β-sheet running along the length of the fibril. ssNMR data indicate that the β-strands are predominately arranged in parallel [114]. What follows are illustrations of some possible elaborations on this basic motif suggested by recent studies on Aβ fibrils [47,48]. (B) Stagger of the β-strands. The stagger, S, can run in either a positive (+1, +2, ...) or negative direction (-1, -2, ...). (C) Register of hydrogen bond (R). A shift in R causes a displacement of one monomer relative to another in the x-y plane. (D) Twist of β-strands within a monomer, T. (E) Packing of multiple β-sheet arrays in the fibril. Two types of symmetries C_{2z} and C_{2x} are shown, which involve a 180° rotation of monomers along the z or x axis, respectively [48]. The two symmetrics predict different polarities for the fibril and hence different mechanisms of elongation.

5.1. High-Resolution Models of Aβ Fibrils from ssNMR

The structure of Aβ(1-40) fibrils [48] based on ssNMR and electron microscopy (EM) consists of a β-hairpin formed by residues 10-23 and 30-36 (Figure 6C). A 180° bend between positions 23-29 brings together the strands. The first 9 residues of Aβ(1-40) are disordered in the fibrils [114]. Along the length of the fibril, neighboring monomers are connected through intermolecular parallel β-sheets (Figure 15A).

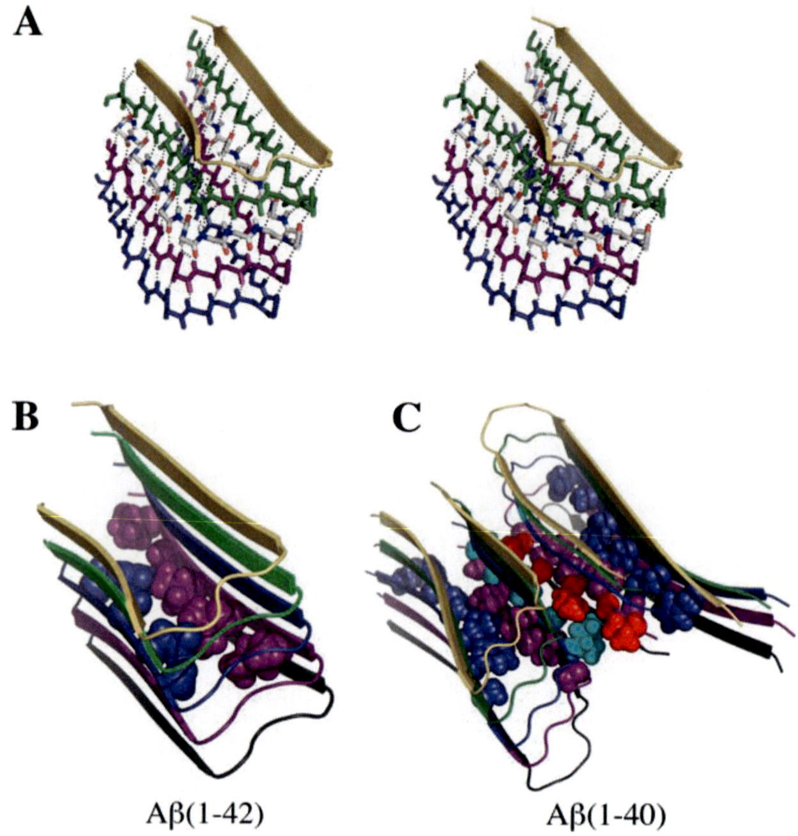

Figure 15. ssNMR structures of Aβ fibrils. (A) Cross-eyed stereo diagram illustrating β-sheet hydrogen bonding in the Aβ (1-42) structure [47]. (B) Illustration of side-chain packing in the Aβ(1-42) structure [47]. Note that each layer of the packing interface is made up from β-strands of adjacent monomers (blue, magenta), with an S=2 stagger. (C) Aβ(1-40) structure [48] illustrating an additional packing interface (red, cyan) between two β-sheet arrays related by C_{2z} symmetry [48].

A second model of the Aβ(1-42) fibril structure (Figure 6D) was proposed based on information from a variety of techniques [47]. X-ray fiber diffraction data were used to infer the overall cross-β arrangement of the strands. The limits of secondary structure were inferred from quenched hydrogen exchange data of the fibrils. Previous ssNMR data [114] on Aβ(1-40) were used to ascertain a parallel intermolecular β-sheet pairing of the Aβ monomers. Site-specific mutagenesis data were used to determine the nature of intra- and intermolecular side-chain packing interactions. Finally, information on the width of the fibrils from electron microscopy was used to constrain quaternary packing. Based on weak protection in qHX experiments, the first 16 residues were modeled as disordered in the Aβ(1-42) structure. By contrast only the first 8 residues are modeled as disordered in the Aβ(1-40) structure [48], consistent with the different qHX properties of these fibrils [122]. The two β-strands in the Aβ(1-42) structure are formed by residues 18-26 and 31-41 (Figure 6D). The 27-30 hairpin turn in the Aβ(1-42) fibrils [47] is thus shorter than the 24-29 turn in the Aβ(1-40)fibrils [48], which is once again consistent with the different qHX behavior of the two types of fibrils

[122]. An Asp23-Lys28 salt-bridge is conserved in both fibrils. Differences in the occupancy of the intramolecular Asp23-Lys28 salt bridge were observed between fibrils grown with or without agitation based on ssNMR, which together with changes in EM fibril morphology illustrate the sensitivity of the fibril structure to subtle changes in growth conditions [48]. Overall the structures of the monomer subunits of Aβ(1-40) and Aβ(1-42) are in good agreement. Significant differences are seen in the lengths of the hairpin turn between the strands, and in the lengths of disordered regions at the ends of the peptide chains. Based on qHX data these differences appear to genuinely reflect differences in the properties of the two fibrils [122].

The quaternary interactions involved in stabilizing the fibril structure have been more difficult to define but significant progress has recently been made in modeling these aspects of the Aβ fibrils structures as well. Because of a paucity of long-range structural constraints an initial model of the Aβ(1-40) fibrils [114] had a number of errors that affected the β-sheet register (and consequently the inside-outside polarity of side chains), as well as the side-chain packing between monomers [47]. This model has now been refined using additional data on quaternary contacts from ssNMR [48]. The most recent Aβ(1-40) [48] and Aβ(1-42) [47] structures provide exquisite detail on the quaternary structures of the fibrils, and with a few exceptions agree quite well with each other.

The repeating unit of the Aβ(1-42) fibril structure consists of a parallel β-sheet dimer comprised of two Aβ peptides (Figure 15B). Side-chains from strand β1 of the first monomer (purple) reach over and pack against the side-chains from strand β2 of the second monomer (blue) [47]. This type of packing is also seen in the revised Aβ(1-40) fibril structure (Figure 15C) and results in an S=2 stagger of β-strands in the nomenclature of the authors [48]. It has been pointed out that the β-strands can stagger with either a +2 or -2 polarity, and that it is currently not possible to distinguish between these alternatives by ssNMR [48].

An important remaining difference between the two Aβ fibril structures is that Aβ(1-42) is modeled as a single β-array per protofilament (Figure 15B), while Aβ(1-40) is modeled as two stacked β-arrays per protofilament (Figure 15C). The differences arise because EM data for the Aβ(1-42) fibrils can be fit with a single molecule per protofilament unit length. For the Aβ(1-40) fibrils, EM and ssNMR data are more consistent with multiple molecules per protofilament unit length (Figure 15C). Differences between the two structures could be due to differences between the Aβ(1-40) and Aβ(1-42) peptides or differences in the conditions used to prepare the fibrils [48]. In this regard it is worth pointing out that the fitting of the structural models to EM data is likely to depend on which segments of the Aβ polypeptide are identified as disordered. The β-strands are ~35 Å long in the Aβ(1-42) structure since residues 1-17 were excluded from calculations. In the Aβ(1-40) structure, where only residues 1-8 were excluded from modeling the strands are ~50 Å long.

Along the width of the protofilament, the monomers in the Aβ(1-40) structure could be stacked (Figure 14E) either such that the free ends of a dimer point in opposite directions (C_{2z} symmetry) or in the same direction (C_{2x} symmetry). Intermolecular distance contacts from ssNMR data are more consistent with C_{2z} symmetry, as is the observation of a single set of side-chain resonances in samples selectively labeled for residues I31, G33, M35, G37, V39 at the interface site. The alternative C_{2x} symmetry [114] would result in non-equivalent magnetic environments of residues from different monomers, and would give rise to a more complex spectrum than observed [48].

Recently the ridges-into-grooves β-sheet packing of a GXXXG motif in Aβ fibrils, which is supported by ssNMR data for the Aβ(1-40) fibril, was exploited to design peptides that act either as inhibitors of fibril formation or to disaggregate fibrils [132]. This study also found that there is a register shift in the sheet-to-sheet packing along the C_{2z} symmetry axis in Aβ(1-40), fibrils compared to Aβ(1-42) fibrils. Presumably this difference is due to the hydrophobic extension in the longer Aβ(1-42) peptide. This difference could explain why Aβ(1-42) aggregates more readily, forms fibrils of greater stability, and is more neurotoxic than the Aβ(1-40) peptide [132].

5.2. Characterization of Trapped Aβ Oligomers by ssNMR

Because evidence is accumulating that toxicity in amyloidogenic diseases is linked more with oligomeric intermediates than with fibrils, there is increasing motivation to obtain a structural characterization of these transiently formed precursors. Encouraging results have recently been obtained by ssNMR [20,119]. First it was demonstrated that oligomeric intermediates accumulate transiently in Aβ(1-40) fibrillization reactions. These soluble species can be separated from the insoluble fibrils by centrifugation. They are active in thioflavin T fluorimetric assays, and have a spherical morphology by EM. The population of intermediates eventually decreases as the mature fibrils start to form. Lowering the temperature of the fibrillization reaction from 37 °C to 4 °C can increase the ratio of intermediates to fibrils. Under certain conditions the intermediates transiently become the predominant species and can be trapped by flash-freezing. Subsequent ssNMR studies can be carried out on the intermediates either as lyophilized powders [20] or in their rehydrated forms [119]. The ssNMR line widths indicate that the intermediates have well-ordered structures. Chemical shift analysis suggest that the Aβ(1-40) monomers adopt very similar structures in the intermediates and in the fibrils [20]. The inter-strand β-sheet distance also appears to be similar based on contacts identified from finite-pulse radiofrequency-driven recoupling experiments. This suggests that both the fibrils and intermediates form in-register parallel β-sheet arrays [20]. In addition to the major form, minor form resonances indicative of polymorphism could be identified. The chemical shifts of some of these minor form resonances were suggestive of α-helical structure [20]. The intermediates isolated in these studies were found to have greater neurotoxicity than the mature fibrils obtained from the same reactions [20]. The intermediates were also shown to be able to 'seed' the formation of mature fibrils, demonstrating they were on-pathway for fibril formation [20].

5.3. ssNMR Studies of αS Fibrils

Although there are currently no high-resolution structural models of αS fibrils, partial ssNMR assignments have established that the polypeptide chain forms a rigid core rich in β-sheet secondary structure, from about residue 22 to 105. The last 35 residues in the C-terminal domain are disordered [133]. The quality of the NMR spectra and structural properties of the fibrils appear to be very sensitive to sample purity and to the conditions under which the fibrils are formed [133,134]. Conversely, once formed the ssNMR spectra of the αS fibrils change little with temperature [135] or upon dehydration [136], suggesting that internal water does not participate in the fibril structure [136]. While this manuscript was being processed, a model for the fold that αS adopts in fibrils was determined based on ssNMR, qHX, and cryo-EM [139]. The αS fragment consisting of residues 30-110, folds into a non-hydrogen bonded 5-stranded antiparallel β-sandwich. Adjacent monomers then connect through intermolecular β-sheet hydrogen bonds running along the length of the fibril [139].

Conclusion

NMR has been at the forefront of amyloid structural biology, and has provided unique insights into the structures of monomeric, membrane-bound, and fibrillar states of amyloidogenic proteins. For monomers, studies have focused on identifying partially folded states that serve as intermediates in protein misfolding. Work on micelle-bound forms is providing the first structural models of amyloidogenic proteins in a membrane environment, which in some cases corresponds to the physiologically and/or toxicologically relevant states. ssNMR has provided structures of amyloid fibrils in unprecedented detail. Although the number of structures is still small, a general model of fibril structures has been proposed. The building block of amyloid fibril structures is a β-sheet. The structure is stabilized in part by side-chain packing, either within a monomer or between monomers. Parallel β-strand pairing between monomers leads to an indefinite hydrogen-bonding network that runs along the length of the fibril. The locations of turns, hinges, and regions of structural disorder show some level of conservation between the different structural states of amyloidogenic proteins.

Future challenges for biophysical work on amyloidogenic proteins include: (1) to determine the extent to which fibril structures vary from the 'general model', particularly in regard to the details of β-sheet packing into higher-order structures. (2) To start addressing the energetic forces responsible for the membrane-bound and fibrillar structures. In this regard, quenched hydrogen exchange studies have already provided some information on solvent accessibilities of hydrogen bonds in fibrils. (3) To make inroads towards the structure determination of oligomeric intermediates that precede fibrils; since these aggregates may be primarily responsible for cytotoxicity. Initial results are encouraging.

Acknowledgements

A.T.A.'s work was supported by a Roger C. Duvoisin research grant from the American Parkinson Disease Association, and in part by NSF grant MCB 0236316. NMR instrumentation was funded in part by grant 1S10RR016760 from the NIH-NCRR. We thank Prof. Peter Lansbury and Dr. Michael J. Volles for the *pT7-7/asyn-WT* expression vector, Dr. Robert Tycko for the coordinates of Aβ(1-40) fibrils, and Prof. Debra Kendall and Arlene Albert for useful discussions about micelles.

References

[1] Chiti, F. and Dobson, C. M. (2006) Protein misfolding, functional amyloid, and human disease. *Annu Rev Biochem. 75*, 333-366

[2] Kelly, J. W. (1998) The alternative conformations of amyloidogenic proteins and their multi-step assembly pathways. *Curr Opin Struct Biol. 8*, 101-106

[3] Koo, E. H., Lansbury, P. T., Jr. and Kelly, J. W. (1999) Amyloid diseases: abnormal protein aggregation in neurodegeneration. *Proc Natl Acad Sci U S A. 96*, 9989-9990

[4] Dobson, C. M. (2003) Protein folding and misfolding. *Nature. 426*, 884-890

[5] Cohen, F. E. and Kelly, J. W. (2003) Therapeutic approaches to protein-misfolding diseases. *Nature. 426*, 905-909

[6] Alexandrescu, A. T. and Rathgeb-Szabo, K. (1999) An NMR investigation of solution aggregation reactions preceding the misassembly of acid-denatured cold shock protein A into fibrils. *J Mol Biol. 291*, 1191-1206

[7] Fandrich, M., Fletcher, M. A. and Dobson, C. M. (2001) Amyloid fibrils from muscle myoglobin. *Nature. 410*, 165-166

[8] Lopez De La Paz, M., Goldie, K., Zurdo, J., Lacroix, E., Dobson, C. M., Hoenger, A. and Serrano, L. (2002) De novo designed peptide-based amyloid fibrils. *Proc Natl Acad Sci U S A. 99*, 16052-16057

[9] Dobson, C. M. (1999) Protein misfolding, evolution and disease. *Trends Biochem Sci. 24*, 329-332

[10] Booth, D. R., Sunde, M., Bellotti, V., Robinson, C. V., Hutchinson, W. L., Fraser, P. E., Hawkins, P. N., Dobson, C. M., Radford, S. E., Blake, C. C. and Pepys, M. B. (1997) Instability, unfolding and aggregation of human lysozyme variants underlying amyloid fibrillogenesis. *Nature. 385*, 787-793

[11] Johnson, S. M., Wiseman, R. L., Sekijima, Y., Green, N. S., Adamski-Werner, S. L. and Kelly, J. W. (2005) Native state kinetic stabilization as a strategy to ameliorate protein misfolding diseases: a focus on the transthyretin amyloidoses. *Acc Chem Res. 38*, 911-921

[12] Riek, R., Wider, G., Billeter, M., Hornemann, S., Glockshuber, R. and Wuthrich, K. (1998) Prion protein NMR structure and familial human spongiform encephalopathies. *Proc Natl Acad Sci U S A. 95*, 11667-11672

[13] Trinh, C. H., Smith, D. P., Kalverda, A. P., Phillips, S. E. and Radford, S. E. (2002) Crystal structure of monomeric human beta-2-microglobulin reveals clues to its amyloidogenic properties. *Proc Natl Acad Sci U S A. 99*, 9771-9776

[14] Eliezer, D., Kutluay, E., Bussell, R., Jr. and Browne, G. (2001) Conformational properties of alpha-synuclein in its free and lipid-associated states. *J Mol Biol. 307*, 1061-1073

[15] Hou, L., Shao, H., Zhang, Y., Li, H., Menon, N. K., Neuhaus, E. B., Brewer, J. M., Byeon, I. J., Ray, D. G., Vitek, M. P., Iwashita, T., Makula, R. A., Przybyla, A. B. and Zagorski, M. G. (2004) Solution NMR studies of the A beta(1-40) and A beta(1-42) peptides establish that the Met35 oxidation state affects the mechanism of amyloid formation. *J Am Chem Soc. 126*, 1992-2005

[16] Williamson, J. A. and Miranker, A. D. (2007) Direct detection of transient alpha-helical states in islet amyloid polypeptide. *Protein Sci. 16*, 110-117

[17] Kanaori, K. and Nosaka, A. Y. (1995) Study of human calcitonin fibrillation by proton nuclear magnetic resonance spectroscopy. *Biochemistry. 34*, 12138-12143

[18] Meyer-Luehmann, M., Coomaraswamy, J., Bolmont, T., Kaeser, S., Schaefer, C., Kilger, E., Neuenschwander, A., Abramowski, D., Frey, P., Jaton, A. L., Vigouret, J. M., Paganetti, P., Walsh, D. M., Mathews, P. M., Ghiso, J., Staufenbiel, M., Walker, L. C. and Jucker, M. (2006) Exogenous induction of cerebral beta-amyloidogenesis is governed by agent and host. *Science. 313*, 1781-1784

[19] Meyer-Luehmann, M., Spires-Jones, T. L., Prada, C., Garcia-Alloza, M., de Calignon, A., Rozkalne, A., Koenigsknecht-Talboo, J., Holtzman, D. M., Bacskai, B. J. and Hyman, B. T. (2008) Rapid appearance and local toxicity of amyloid-beta plaques in a mouse model of Alzheimer's disease. *Nature. 451*, 720-724

[20] Chimon, S., Shaibat, M. A., Jones, C. R., Calero, D. C., Aizezi, B. and Ishii, Y. (2007) Evidence of fibril-like beta-sheet structures in a neurotoxic amyloid intermediate of Alzheimer's beta-amyloid. *Nat Struct Mol Biol. 14,* 1157-1164

[21] Ferreira, S. T., Vieira, M. N. and De Felice, F. G. (2007) Soluble protein oligomers as emerging toxins in Alzheimer's and other amyloid diseases. IUBMB Life. *59*, 332-345

[22] Lansbury, P. T. and Lashuel, H. A. (2006) A century-old debate on protein aggregation and neurodegeneration enters the clinic. *Nature. 443*, 774-779

[23] Ross, C. A. and Poirier, M. A. (2004) Protein aggregation and neurodegenerative disease. *Nat Med. 10 Suppl*, S10-17

[24] Donahue, J. E., Berzin, T. M., Rafii, M. S., Glass, D. J., Yancopoulos, G. D., Fallon, J. R. and Stopa, E. G. (1999) Agrin in Alzheimer's disease: altered solubility and abnormal distribution within microvasculature and brain parenchyma. *Proc Natl Acad Sci U S A. 96*, 6468-6472

[25] Bush, A. I. (2003) The metallobiology of Alzheimer's disease. *Trends Neurosci. 26*, 207-214

[26] Zhang, Q., Powers, E. T., Nieva, J., Huff, M. E., Dendle, M. A., Bieschke, J., Glabe, C. G., Eschenmoser, A., Wentworth, P., Jr., Lerner, R. A. and Kelly, J. W. (2004) Metabolite-initiated protein misfolding may trigger Alzheimer's disease. *Proc Natl Acad Sci U S A. 101*, 4752-4757

[27] Lashuel, H. A., Hartley, D., Petre, B. M., Walz, T. and Lansbury, P. T., Jr. (2002) Neurodegenerative disease: amyloid pores from pathogenic mutations. Nature. *418*, 291

[28] Ernst, R. L. and Hay, J. W. (1994) The US economic and social costs of Alzheimer's disease revisited. *Am J Public Health. 84*, 1261-1264

[29] Bush, A. I. (2002) Metal complexing agents as therapies for Alzheimer's disease. *Neurobiol Aging. 23*, 1031-1038

[30] Cotman, S. L., Halfter, W. and Cole, G. J. (2000) Agrin binds to beta-amyloid (Abeta), accelerates abeta fibril formation, and is localized to Abeta deposits in Alzheimer's disease brain. *Mol Cell Neurosci. 15*, 183-198

[31] Selkoe, D. J. (2001) Presenilin, Notch, and the genesis and treatment of Alzheimer's disease. *Proc Natl Acad Sci U S A. 98*, 11039-11041

[32] Kayed, R., Head, E., Thompson, J. L., McIntire, T. M., Milton, S. C., Cotman, C. W. and Glabe, C. G. (2003) Common structure of soluble amyloid oligomers implies common mechanism of pathogenesis. *Science. 300*, 486-489

[33] Lambert, M. P., Barlow, A. K., Chromy, B. A., Edwards, C., Freed, R., Liosatos, M., Morgan, T. E., Rozovsky, I., Trommer, B., Viola, K. L., Wals, P., Zhang, C., Finch, C. E., Krafft, G. A. and Klein, W. L. (1998) Diffusible, nonfibrillar ligands derived from Abeta1-42 are potent central nervous system neurotoxins. *Proc Natl Acad Sci U S A. 95*, 6448-6453

[34] George, J. M. (2002) The synucleins. *Genome Biol. 3*, REVIEWS3002

[35] Cookson, M. R. (2005) The biochemistry of Parkinson's disease. *Annu Rev Biochem. 74*, 29-52

[36] Moore, D. J., West, A. B., Dawson, V. L. and Dawson, T. M. (2005) Molecular pathophysiology of Parkinson's disease. *Annu Rev Neurosci. 28*, 57-87

[37] Goedert, M. (2001) Alpha-synuclein and neurodegenerative diseases. *Nat Rev Neurosci. 2*, 492-501

[38] Fink, A. L. (2006) The aggregation and fibrillation of alpha-synuclein. *Acc Chem Res. 39*, 628-634

[39] Conway, K. A., Lee, S. J., Rochet, J. C., Ding, T. T., Williamson, R. E. and Lansbury, P. T., Jr. (2000) Acceleration of oligomerization, not fibrillization, is a shared property of both alpha-synuclein mutations linked to early-onset Parkinson's disease: implications for pathogenesis and therapy. *Proc Natl Acad Sci U S A. 97*, 571-576

[40] El-Agnaf, O. M., Salem, S. A., Paleologou, K. E., Cooper, L. J., Fullwood, N. J., Gibson, M. J., Curran, M. D., Court, J. A., Mann, D. M., Ikeda, S., Cookson, M. R., Hardy, J. and Allsop, D. (2003) Alpha-synuclein implicated in Parkinson's disease is present in extracellular biological fluids, including human plasma. *Faseb J. 17*, 1945-1947

[41] Weinreb, P. H., Zhen, W., Poon, A. W., Conway, K. A. and Lansbury, P. T., Jr. (1996) NACP, a protein implicated in Alzheimer's disease and learning, is natively unfolded. *Biochemistry. 35*, 13709-13715

[42] Wuthrich, K. (2003) NMR studies of structure and function of biological macromolecules (Nobel lecture). *Angew Chem Int Ed Engl. 42*, 3340-3363

[43] Shortle, D. R. (1996) Structural analysis of non-native states of proteins by NMR methods. *Curr Opin Struct Biol. 6*, 24-30

[44] Alexandrescu, A. T. (2001) An NMR-based quenched hydrogen exchange investigation of model amyloid fibrils formed by cold shock protein A. *Pac Symp Biocomput*, 67-78

[45] McDermott, A. E. (2004) Structural and dynamic studies of proteins by solid-state NMR spectroscopy: rapid movement forward. *Curr Opin Struct Biol. 14*, 554-561

[46] Tycko, R. (2006) Molecular structure of amyloid fibrils: insights from solid-state NMR. *Q Rev Biophys. 39*, 1-55

[47] Luhrs, T., Ritter, C., Adrian, M., Riek-Loher, D., Bohrmann, B., Dobeli, H., Schubert, D. and Riek, R. (2005) 3D structure of Alzheimer's amyloid-beta(1-42) fibrils. *Proc Natl Acad Sci U S A. 102*, 17342-17347

[48] Petkova, A. T., Yau, W. M. and Tycko, R. (2006) Experimental constraints on quaternary structure in Alzheimer's beta-amyloid fibrils. *Biochemistry. 45*, 498-512

[49] Ferguson, N., Becker, J., Tidow, H., Tremmel, S., Sharpe, T. D., Krause, G., Flinders, J., Petrovich, M., Berriman, J., Oschkinat, H. and Fersht, A. R. (2006) General structural motifs of amyloid protofilaments. *Proc Natl Acad Sci U S A. 103*, 16248-16253

[50] Zhang, J., Yarowsky, P., Gordon, M. N., Di Carlo, G., Munireddy, S., van Zijl, P. C. and Mori, S. (2004) Detection of amyloid plaques in mouse models of Alzheimer's disease by magnetic resonance imaging. *Magn Reson Med. 51*, 452-457

[51] Kiss, E., Keusch, G., Zanetti, M., Jung, T., Schwarz, A., Schocke, M., Jaschke, W. and Czermak, B. V. (2005) Dialysis-related amyloidosis revisited. *AJR Am J Roentgenol. 185*, 1460-1467

[52] Whitwell, J. L. and Jack, C. R., Jr. (2007) Neuroimaging in dementia. *Neurol Clin. 25*, 843-857, viii

[53] Riek, R., Hornemann, S., Wider, G., Billeter, M., Glockshuber, R. and Wuthrich, K. (1996) NMR structure of the mouse prion protein domain PrP(121-321). *Nature. 382*, 180-182

[54] Cobb, N. J., Sonnichsen, F. D., McHaourab, H. and Surewicz, W. K. (2007) Molecular architecture of human prion protein amyloid: a parallel, in-register beta-structure. *Proc Natl Acad Sci U S A. 104*, 18946-18951

[55] Bai, Y., Milne, J. S., Mayne, L. and Englander, S. W. (1994) Protein stability parameters measured by hydrogen exchange. *Proteins. 20*, 4-14

[56] Dempsey, C. E. (2001) Hydrogen exchange in peptides and proteins using NMR spectroscopy. *Progress in Nuclear Magnetic Resonance Spectroscopy. 39*, 135-170

[57] Liu, K., Kelly, J. W. and Wemmer, D. E. (2002) Native state hydrogen exchange study of suppressor and pathogenic variants of transthyretin. *J Mol Biol. 320*, 821-832

[58] Dumoulin, M., Canet, D., Last, A. M., Pardon, E., Archer, D. B., Muyldermans, S., Wyns, L., Matagne, A., Robinson, C. V., Redfield, C. and Dobson, C. M. (2005) Reduced global cooperativity is a common feature underlying the amyloidogenicity of pathogenic lysozyme mutations. *J Mol Biol. 346*, 773-788

[59] Liu, H., Farr-Jones, S., Ulyanov, N. B., Llinas, M., Marqusee, S., Groth, D., Cohen, F. E., Prusiner, S. B. and James, T. L. (1999) Solution structure of Syrian hamster prion protein rPrP(90-231*). Biochemistry. 38*, 5362-5377

[60] Shipp, E. L., Cantini, F., Bertini, I., Valentine, J. S. and Banci, L. (2003) Dynamic properties of the G93A mutant of copper-zinc superoxide dismutase as detected by NMR spectroscopy: implications for the pathology of familial amyotrophic lateral sclerosis. *Biochemistry. 42*, 1890-1899

[61] Palmer, A. G., 3rd. (2001) Nmr probes of molecular dynamics: overview and comparison with other techniques. *Annu Rev Biophys Biomol Struct. 30*, 129-155

[62] Alexandrescu, A. T., Jahnke, W., Wiltscheck, R. and Blommers, M. J. (1996) Accretion of structure in staphylococcal nuclease: an 15N NMR relaxation study. *J Mol Biol. 260*, 570-587

[63] Wang, Y. and Shortle, D. (1996) A dynamic bundle of four adjacent hydrophobic segments in the denatured state of staphylococcal nuclease. *Protein Sci. 5*, 1898-1906

[64] Alexandrescu, A. T., Lamour, F. P. and Jaravine, V. A. (2000) NMR evidence for progressive stabilization of native-like structure upon aggregation of acid-denatured LysN. J Mol Biol. *295*, 239-255

[65] Croke, R. L., Sallum, C. O., Watson, E. and Alexandrescu, A. T. (2008) Hydrogen exchange of monomeric α-synuclein shows the unfolded structure is retained at physiological temperature and is independent of molecular crowding in Escherichia coli. *Protein Sci*., *in press.*

[66] McParland, V. J., Kalverda, A. P., Homans, S. W. and Radford, S. E. (2002) Structural properties of an amyloid precursor of beta(2)-microglobulin. *Nat Struct Biol. 9*, 326-331

[67] Hoshino, M., Katou, H., Hagihara, Y., Hasegawa, K., Naiki, H. and Goto, Y. (2002) Mapping the core of the beta(2)-microglobulin amyloid fibril by H/D exchange. *Nat Struct Biol. 9*, 332-336

[68] Alexandrescu, A. T., Abeygunawardana, C. and Shortle, D. (1994) Structure and dynamics of a denatured 131-residue fragment of staphylococcal nuclease: a heteronuclear NMR study. *Biochemistry. 33*, 1063-1072

[69] Riek, R., Guntert, P., Dobeli, H., Wipf, B. and Wuthrich, K. (2001) NMR studies in aqueous solution fail to identify significant conformational differences between the monomeric forms of two Alzheimer peptides with widely different plaque-competence, A beta(1-40)(ox) and A beta(1-42)(ox). *Eur J Biochem. 268*, 5930-5936

[70] Eliezer, D., Barre, P., Kobaslija, M., Chan, D., Li, X. and Heend, L. (2005) Residual structure in the repeat domain of tau: echoes of microtubule binding and paired helical filament formation. *Biochemistry. 44*, 1026-1036

[71] Mukrasch, M. D., Biernat, J., von Bergen, M., Griesinger, C., Mandelkow, E. and Zweckstetter, M. (2005) Sites of tau important for aggregation populate {beta}-structure and bind to microtubules and polyanions. *J Biol Chem. 280*, 24978-24986

[72] Watson, A. A., Fairlie, D. P. and Craik, D. J. (1998) Solution structure of methionine-oxidized amyloid beta-peptide (1-40). Does oxidation affect conformational switching? *Biochemistry. 37*, 12700-12706

[73] Louhivuori, M., Paakkonen, K., Fredriksson, K., Permi, P., Lounila, J. and Annila, A. (2003) On the origin of residual dipolar couplings from denatured proteins. *J Am Chem Soc. 125*, 15647-15650

[74] Danielsson, J., Andersson, A., Jarvet, J. and Graslund, A. (2006) 15N relaxation study of the amyloid beta-peptide: structural propensities and persistence length. *Magn Reson Chem. 44 Spec No*, S114-121

[75] Yan, Y. and Wang, C. (2006) Abeta42 is more rigid than Abeta40 at the C terminus: implications for Abeta aggregation and toxicity. *J Mol Biol. 364*, 853-862

[76] Danielsson, J., Pierattelli, R., Banci, L. and Graslund, A. (2007) High-resolution NMR studies of the zinc-binding site of the Alzheimer's amyloid beta-peptide. *FEBS J. 274*, 46-59

[77] Hou, L. and Zagorski, M. G. (2006) NMR reveals anomalous copper(II) binding to the amyloid Abeta peptide of Alzheimer's disease. *J Am Chem Soc. 128*, 9260-9261

[78] Williamson, M. P., Suzuki, Y., Bourne, N. T. and Asakura, T. (2006) Binding of amyloid beta-peptide to ganglioside micelles is dependent on histidine-13. *Biochem J. 397*, 483-490

[79] Petkova, A. T., Buntkowsky, G., Dyda, F., Leapman, R. D., Yau, W. M. and Tycko, R. (2004) Solid state NMR reveals a pH-dependent antiparallel beta-sheet registry in fibrils formed by a beta-amyloid peptide. *J Mol Biol. 335*, 247-260

[80] Lazo, N. D., Grant, M. A., Condron, M. C., Rigby, A. C. and Teplow, D. B. (2005) On the nucleation of amyloid beta-protein monomer folding. *Protein Sci. 14*, 1581-1596

[81] McLaurin, J., Franklin, T., Zhang, X., Deng, J. and Fraser, P. E. (1999) Interactions of Alzheimer amyloid-beta peptides with glycosaminoglycans effects on fibril nucleation and growth. *Eur J Biochem. 266*, 1101-1110

[82] Yan, Y. and Wang, C. (2007) Abeta40 protects non-toxic Abeta42 monomer from aggregation. *J Mol Biol. 369*, 909-916

[83] Narayanan, S. and Reif, B. (2005) Characterization of chemical exchange between soluble and aggregated states of beta-amyloid by solution-state NMR upon variation of salt conditions. *Biochemistry. 44*, 1444-1452

[84] McNulty, B. C., Tripathy, A., Young, G. B., Charlton, L. M., Orans, J. and Pielak, G. J. (2006) Temperature-induced reversible conformational change in the first 100 residues of alpha-synuclein. *Protein Sci. 15*, 602-608

[85] Bertoncini, C. W., Jung, Y. S., Fernandez, C. O., Hoyer, W., Griesinger, C., Jovin, T. M. and Zweckstetter, M. (2005) Release of long-range tertiary interactions potentiates aggregation of natively unstructured alpha-synuclein. *Proc Natl Acad Sci U S A. 102*, 1430-1435

[86] Bussell, R., Jr. and Eliezer, D. (2001) Residual structure and dynamics in Parkinson's disease-associated mutants of alpha-synuclein. *J Biol Chem. 276*, 45996-46003

[87] Bernado, P., Bertoncini, C. W., Griesinger, C., Zweckstetter, M. and Blackledge, M. (2005) Defining long-range order and local disorder in native alpha-synuclein using residual dipolar couplings. *J Am Chem Soc. 127*, 17968-17969

[88] Cho, M. K., Kim, H. Y., Bernado, P., Fernandez, C. O., Blackledge, M. and Zweckstetter, M. (2007) Amino acid bulkiness defines the local conformations and dynamics of natively unfolded alpha-synuclein and tau. J Am Chem Soc. *129*, 3032-3033

[89] Sung, Y. H. and Eliezer, D. (2007) Residual structure, backbone dynamics, and interactions within the synuclein family. *J Mol Biol. 372*, 689-707

[90] Dedmon, M. M., Lindorff-Larsen, K., Christodoulou, J., Vendruscolo, M. and Dobson, C. M. (2005) Mapping long-range interactions in alpha-synuclein using spin-label NMR and ensemble molecular dynamics simulations. *J Am Chem Soc. 127*, 476-477

[91] Clore, G. M., Tang, C. and Iwahara, J. (2007) Elucidating transient macromolecular interactions using paramagnetic relaxation enhancement. *Curr Opin Struct Biol. 17*, 603-616

[92] Fernandez, C. O., Hoyer, W., Zweckstetter, M., Jares-Erijman, E. A., Subramaniam, V., Griesinger, C. and Jovin, T. M. (2004) NMR of alpha-synuclein-polyamine complexes elucidates the mechanism and kinetics of induced aggregation. *EMBO J. 23*, 2039-2046

[93] McNulty, B. C., Young, G. B. and Pielak, G. J. (2006) Macromolecular crowding in the Escherichia coli periplasm maintains alpha-synuclein disorder. *J Mol Bi*ol. *355*, 893-897

[94] Marsh, J. A., Singh, V. K., Jia, Z. and Forman-Kay, J. D. (2006) Sensitivity of secondary structure propensities to sequence differences between alpha- and gamma-synuclein: implications for fibrillation. *Protein Sci. 15*, 2795-2804

[95] Hwang, T. L., van Zijl, P. C. and Mori, S. (1998) Accurate quantitation of water-amide proton exchange rates using the phase-modulated CLEAN chemical EXchange (CLEANEX-PM) approach with a Fast-HSQC (FHSQC) detection scheme. *J Biomol NMR. 11*, 221-226

[96] Shimba, N., Serber, Z., Ledwidge, R., Miller, S. M., Craik, C. S. and Dotsch, V. (2003) Quantitative identification of the protonation state of histidines in vitro and in vivo. *Biochemistry. 42*, 9227-9234

[97] Rasia, R. M., Bertoncini, C. W., Marsh, D., Hoyer, W., Cherny, D., Zweckstetter, M., Griesinger, C., Jovin, T. M. and Fernandez, C. O. (2005) Structural characterization of copper(II) binding to alpha-synuclein: Insights into the bioinorganic chemistry of Parkinson's disease. *Proc Natl Acad Sci U S A. 102*, 4294-4299

[98] Sung, Y. H., Rospigliosi, C. and Eliezer, D. (2006) NMR mapping of copper binding sites in alpha-synuclein. *Biochim Biophys Acta. 1764*, 5-12

[99] Binolfi, A., Rasia, R. M., Bertoncini, C. W., Ceolin, M., Zweckstetter, M., Griesinger, C., Jovin, T. M. and Fernandez, C. O. (2006) Interaction of alpha-synuclein with divalent metal ions reveals key differences: a link between structure, binding specificity and fibrillation enhancement. *J Am Chem Soc. 128*, 9893-9901

[100] Mandal, P. K., Pettegrew, J. W., Masliah, E., Hamilton, R. L. and Mandal, R. (2006) Interaction between Abeta peptide and alpha synuclein: molecular mechanisms in overlapping pathology of Alzheimer's and Parkinson's in dementia with Lewy body disease. *Neurochem Res. 31*, 1153-1162

[101] Mandal, P. K. and Pettegrew, J. W. (2007) "Interaction between Abeta Peptide and alpha Synuclein: Molecular Mechanisms in Overlapping Pathology of Alzheimer's and Parkinson's in Dementia with Lewy Body Disease". *Neurochem Res 33,* 220

[102] Fredenburg, R. A., Rospigliosi, C., Meray, R. K., Kessler, J. C., Lashuel, H. A., Eliezer, D. and Lansbury, P. T., Jr. (2007) The impact of the E46K mutation on the properties of alpha-synuclein in its monomeric and oligomeric states. *Biochemistry. 46*, 7107-7118

[103] Bertoncini, C. W., Rasia, R. M., Lamberto, G. R., Binolfi, A., Zweckstetter, M., Griesinger, C. and Fernandez, C. O. (2007) Structural characterization of the intrinsically unfolded protein beta-synuclein, a natural negative regulator of alpha-synuclein aggregation. *J Mol Biol. 372*, 708-722

[104] Ulmer, T. S., Bax, A., Cole, N. B. and Nussbaum, R. L. (2005) Structure and dynamics of micelle-bound human alpha-synuclein. *J Biol Chem. 280*, 9595-9603

[105] Parker, W. and Song, P. S. (1992) Protein structures in SDS micelle-protein complexes. *Biophys J. 61*, 1435-1439

[106] Chandra, S., Chen, X., Rizo, J., Jahn, R. and Sudhof, T. C. (2003) A broken alpha - helix in folded alpha -Synuclein. *J Biol Chem. 278*, 15313-15318

[107] Jao, C. C., Der-Sarkissian, A., Chen, J. and Langen, R. (2004) Structure of membrane-bound alpha-synuclein studied by site-directed spin labeling. *Proc Natl Acad Sci U S A. 101*, 8331-8336

[108] Amodeo, P., Motta, A., Strazzullo, G. and Castiglione Morelli, M. A. (1999) Conformational flexibility in calcitonin: the dynamic properties of human and salmon calcitonin in solution. *J Biomol NMR. 13*, 161-174

[109] Barre, P. and Eliezer, D. (2006) Folding of the repeat domain of tau upon binding to lipid surfaces. *J Mol Biol. 362*, 312-326

[110] Jarvet, J., Danielsson, J., Damberg, P., Oleszczuk, M. and Graslund, A. (2007) Positioning of the Alzheimer Abeta(1-40) peptide in SDS micelles using NMR and paramagnetic probes. *J Biomol NMR. 39*, 63-72

[111] Shao, H., Jao, S., Ma, K. and Zagorski, M. G. (1999) Solution structures of micelle-bound amyloid beta-(1-40) and beta-(1-42) peptides of Alzheimer's disease. *J Mol Biol. 285*, 755-773

[112] Coles, M., Bicknell, W., Watson, A. A., Fairlie, D. P. and Craik, D. J. (1998) Solution structure of amyloid beta-peptide(1-40) in a water-micelle environment. Is the membrane-spanning domain where we think it is? *Biochemistry. 37*, 11064-11077

[113] Mandal, P. K. and Pettegrew, J. W. (2004) Alzheimer's disease: NMR studies of asialo (GM1) and trisialo (GT1b) ganglioside interactions with Abeta(1-40) peptide in a membrane mimic environment. *Neurochem Res. 29*, 447-453

[114] Petkova, A. T., Ishii, Y., Balbach, J. J., Antzutkin, O. N., Leapman, R. D., Delaglio, F. and Tycko, R. (2002) A structural model for Alzheimer's beta -amyloid fibrils based on experimental constraints from solid state NMR. *Proc Natl Acad Sci U S A. 99*, 16742-16747

[115] Marcelino, A. M. and Gierasch, L. M. (2008) Roles of beta-turns in protein folding: From peptide models to protein engineering. *Biopolymers. 89*, 380-391

[116] Ulmer, T. S. and Bax, A. (2005) Comparison of structure and dynamics of micelle-bound human alpha-synuclein and Parkinson disease variants. *J Biol Chem. 280*, 43179-43187

[117] Chen, Z., Krause, G. and Reif, B. (2005) Structure and orientation of peptide inhibitors bound to beta-amyloid fibrils. *J Mol B*iol. *354*, 760-776

[118] Zhang, Y. Z., Paterson, Y. and Roder, H. (1995) Rapid amide proton exchange rates in peptides and proteins measured by solvent quenching and two-dimensional NMR. *Protein Sci. 4*, 804-814

[119] Chimon, S. and Ishii, Y. (2005) Capturing intermediate structures of Alzheimer's beta-amyloid, Abeta(1-40), by solid-state NMR spectroscopy. *J Am Chem Soc. 127*, 13472-13473

[120] Morgan, G. J., Giannini, S., Hounslow, A. M., Craven, C. J., Zerovnik, E., Turk, V., Waltho, J. P. and Staniforth, R. A. (2008) Exclusion of the native alpha-helix from the amyloid fibrils of a mixed alpha/beta protein. *J Mol Biol. 375*, 487-498

[121] Whittemore, N. A., Mishra, R., Kheterpal, I., Williams, A. D., Wetzel, R. and Serpersu, E. H. (2005) Hydrogen-deuterium (H/D) exchange mapping of Abeta 1-40 amyloid fibril secondary structure using nuclear magnetic resonance spectroscopy. *Biochemistry. 44*, 4434-4441

[122] Olofsson, A., Lindhagen-Persson, M., Sauer-Eriksson, A. E. and Ohman, A. (2007) Amide solvent protection analysis demonstrates that amyloid-beta(1-40) and amyloid-beta(1-42) form different fibrillar structures under identical conditions. *Biochem J. 404*, 63-70

[123] Olofsson, A., Sauer-Eriksson, A. E. and Ohman, A. (2006) The solvent protection of alzheimer amyloid-beta-(1-42) fibrils as determined by solution NMR spectroscopy. *J Biol Chem. 281*, 477-483

[124] Wilson, L. M., Mok, Y. F., Binger, K. J., Griffin, M. D., Mertens, H. D., Lin, F., Wade, J. D., Gooley, P. R. and Howlett, G. J. (2007) A structural core within apolipoprotein C-II amyloid fibrils identified using hydrogen exchange and proteolysis. *J Mol Biol. 366*, 1639-1651

[125] Yamaguchi, K., Katou, H., Hoshino, M., Hasegawa, K., Naiki, H. and Goto, Y. (2004) Core and heterogeneity of beta2-microglobulin amyloid fibrils as revealed by H/D exchange. *J Mol Biol. 338*, 559-571

[126] Olofsson, A., Ippel, J. H., Wijmenga, S. S., Lundgren, E. and Ohman, A. (2004) Probing solvent accessibility of transthyretin amyloid by solution NMR spectroscopy. *J Biol Chem. 279*, 5699-5707

[127] Kheterpal, I. and Wetzel, R. (2006) Hydrogen/deuterium exchange mass spectrometry--a window into amyloid structure. *Acc Chem Res. 39*, 584-593

[128] Alexandrescu, A. T. (2005) Amyloid accomplices and enforcers. *Protein Sci. 14*, 1-12

[129] Bai, Y., Sosnick, T. R., Mayne, L. and Englander, S. W. (1995) Protein folding intermediates: native-state hydrogen exchange. *Science. 269*, 192-197.

[130] Tycko, R. (2006) Characterization of amyloid structures at the molecular level by solid state nuclear magnetic resonance spectroscopy. *Methods Enzymol. 413*, 103-122

[131] Iwata, K., Fujiwara, T., Matsuki, Y., Akutsu, H., Takahashi, S., Naiki, H. and Goto, Y. (2006) 3D structure of amyloid protofilaments of beta2-microglobulin fragment probed by solid-state NMR. *Proc Natl Acad Sci U S* A. *103*, 18119-18124

[132] Sato, T., Kienlen-Campard, P., Ahmed, M., Liu, W., Li, H., Elliott, J. I., Aimoto, S., Constantinescu, S. N., Octave, J. N. and Smith, S. O. (2006) Inhibitors of amyloid toxicity based on beta-sheet packing of Abeta40 and Abeta42. *Biochemistry. 45*, 5503-5516

[133] Heise, H., Hoyer, W., Becker, S., Andronesi, O. C., Riedel, D. and Baldus, M. (2005) Molecular-level secondary structure, polymorphism, and dynamics of full-length alpha-synuclein fibrils studied by solid-state NMR. *Proc Natl Acad Sci U S A. 102*, 15871-15876

[134] Kloepper, K. D., Woods, W. S., Winter, K. A., George, J. M. and Rienstra, C. M. (2006) Preparation of alpha-synuclein fibrils for solid-state NMR: expression, purification, and incubation of wild-type and mutant forms. *Protein Expr Purif. 48*, 112-117

[135] Kloepper, K. D., Zhou, D. H., Li, Y., Winter, K. A., George, J. M. and Rienstra, C. M. (2007) Temperature-dependent sensitivity enhancement of solid-state NMR spectra of alpha-synuclein fibrils. *J Biomol NMR. 39*, 197-211

[136] Kloepper, K. D., Hartman, K. L., Ladror, D. T. and Rienstra, C. M. (2007) Solid-state NMR spectroscopy reveals that water is nonessential to the core structure of alpha-synuclein fibrils. *J Phys Chem B. 111*, 13353-13356

[137] Wishart, D. S., Bigam, C. G., Holm, A., Hodges, R. S. and Sykes, B. D. (1995) 1H, 13C and 15N random coil NMR chemical shifts of the common amino acids. I. Investigations of nearest-neighbor effects. *J Biomol NMR. 5*, 67-81

[138] Tomaselli, S., Esposito, V., Vangone, P., van Nuland, N. A., Bonvin, A. M., Guerrini, R., Tancredi, T., Temussi, P. A. and Picone, D. (2006) The alpha-to-beta conformational transition of Alzheimer's Abeta-(1-42) peptide in aqueous media is reversible: a step by step conformational analysis suggests the location of beta conformation seeding. *Chembiochem. 7*, 257-2673.

[139] Vilar, M., Chou, H-T., Luhrs, T. Maji, S.K., Riek-Loher, D., Verel, R., Manning, G., Stahlberg, H., and Riek, R. (2008) The fold of α-synuclein fibrils. *Proc Natl Acad Sci U S A. 105*, 8637-8642.

In: Protein Misfolding
Editors: C. B. O'Doherty, A.C. Byrne

ISBN: 978-1-60456-881-3

Chapter II

PROTEIN MISFOLDING: A FRAMEWORK FOR THE EMERGENCE OF NOVEL BIOLOGICAL FUNCTIONS*

David A. Rozak[1], Wes Swietnicki[1] and Philip N. Bryan[2]
[1]Bacteriology Division
United States Army Medical Research Institute for Infectious Diseases, USA
[2]Center for Advanced Research in Biotechnology
University of Maryland Biotechnology Institute, USA

ABSTRACT

Proteins occasionally exhibit the potential to adopt alternate folds under natural or artificial conditions. These conformational changes can lead to the creation of disease-forming amyloid plaques or accumulation of insoluble aggregates in over-expressed recombinant proteins. Because alternate folds challenge the traditional notion of structurally defined proteins and frequently lead to problems in natural and laboratory settings, it is tempting to characterize structurally ambiguous proteins as ill-behaved. Unfortunately, this perspective masks the positive role structural ambiguity can play in protein function and evolution. With an estimated 30% of eukaryotic proteins containing extensive disordered regions and mutagenesis experiments showing that a minimal number of polymorphisms are required to specify dramatically different structures, protein instability and conformational switching may play a greater role in cell biology than previously expected. We review an expanding body of research that encourages a more relaxed interpretation of the central dogma and promotes a model for protein evolution in which conformational switching uncovers novel function by exposing cryptic epitopes. Under this model, prion-induced diseases represent a negative outcome

* The views of the authors do not purport to reflect the positions of the Department of the Army or the Department of Defense.

of a natural process that helps organisms explore available sequence space for positive biological activity.

INTRODUCTION

The central dogma of molecular biology seeks to establish a clear and unambiguous link between genotype and phenotype through a series of steps involving genetic translation and transcription, protein folding, and physiological function. A core principle of this dogma is that a protein's amino acid sequence is sufficient to reproducibly define its three-dimensional structure and that this structure directly determines how the molecule will behave in the organism. Because the protein's conformation generally supports its ligand binding and enzymatic activities, folding is often seen as a necessary step in the fulfillment of the genotype-phenotype link.

The current concept of protein misfolding is in part based on the understanding that there is an optimal, well-defined, conformation for proteins to assume and that there exists a class of ill-behaved proteins, which have the propensity to fold in manners that are problematic for the organism and/or the biologist. Prions have come to represent the classic example of misfolded proteins which adversely affect the organism, thereby causing disease. First described by Stanley Prusiner in 1982 (Prusiner 1982), the membrane-bound proteins have the propensity to form multimeric plaques via self-propagating alpha-beta switches (Pan, Baldwin et al. 1993). The disease-causing plaques are disruptive to cell function and lead to severe neurological disorders in humans and other animals (Prusiner, Scott et al. 1998). In addition to prions and similar plaque-forming proteins (Carrell and Gooptu 1998), biologists routinely struggle with misfolded proteins in the lab. Here, protein misfolding is generally caused by efforts to over express the proteins in foreign cell types under conditions that do not represent natural environments. However, prions and other ill-behaved proteins are not unique in their ability to adopt alternate folds.

Indeed, a broad array of peer-reviewed studies indicates that the protein code is frequently insufficient to specify structure and function. In many cases, ordered and disordered protein elements are induced to adopt alternate conformations by environmental factors such as temperature, pH, and ligand interactions. This has lead researchers to challenge the classic notion that structure is a prerequisite for protein function. The main assault on the view that function follows form is being waged by bioinformaticians who have shown that as many as one third of all eukaryotic proteins contain extensive disordered regions under physiological conditions and that these regions have distinct biological functions. Experimental data indicate that, in many cases, disordered regions adopt induced-folds only after binding their cognate ligands. Studies by several groups also indicate that the protein code first proposed by Anfinsen (Anfinsen, Haber et al. 1961; Anfinsen 1973) and subsequently reflected in an array of sequence-based structural prediction models (Schiffer and Edmundson 1967; Pain and Robson 1970; Chou and Fasman 1978; Levin, Robson et al. 1986; Nishikawa and Ooi 1986; Sweet 1986; Biou, Gibrat et al. 1988; Bohr, Bohr et al. 1988; Qian and Sejnowski 1988; Holley and Karplus 1989) are often ambiguous and susceptible to influence by non-local forces. In many cases, global or environmental cues are sufficient to override the structural propensity of local sequence content. A careful review of these data suggests biologists might benefit from a refined understanding of the central dogma, which reflects the ambiguities observed in the protein code and no longer holds structure as a prerequisite for function. The fact that molten and disordered protein elements exhibit distinct biological functions has important ramifications for the evolution of novel biological function, which we will explore at the end of this chapter.

Eukaryotic Proteins Frequently Contain Extensive Disordered Regions

Biologists have traditionally described proteins as being composed of locally defined structural components such as alpha helices and beta folds, which interact with distal regions to render more complex tertiary structures and quaternary structures. For the most part, these well-defined structures are thought to be formed shortly after the nascent protein emerges from the ribosome. Although some degree of motion is expected and frequently observed during the molecule's existence, a well-defined structure is generally thought to be a prerequisite for protein interactions and enzymatic activity. The traditional view that protein structure enables function may partially result from the biases and limits inherent in our laboratory technologies.

Given the experimental requirements of X-ray crystallography and nuclear magnetic resonance spectroscopy, it is not surprising that structural studies are biased toward nicely folded proteins, which reinforce the elegant but somewhat simplistic expectations of a well-ordered proteome. Indeed, a review of Protein Data Bank (PDB) structures has revealed that low-complexity sequences, similar to those associated with disordered regions, are significantly under-represented in the PDB structural database compared to other sequence data sets (Huntley and Golding 2002). Nevertheless, it is clear from some of these structures that even well-ordered proteins contain extensive disordered regions, which appear as conspicuous gaps in the solution structures (Wyatt, Kwong et al. 1998; Wright and Dyson 1999; Uversky, Gillespie et al. 2000; Okada, Hirotsu et al. 2001; Weiss 2001; Yaremchuk, Tukalo et al. 2001).

Our appreciation for the biological diversity and function of unstructured regions may also be limited by spectroscopic techniques, which generally cannot differentiate proteins that are predominantly in the unfolded form. For example, a protein for which 1 out of 100 molecules occupy the folded state ($\Delta G_{unfolding}$ = -2.7 kcal/mol) will appear to be similar to a protein in which the folded state is populated by 1 out of 10^7 molecules ($\Delta G_{unfolding}$ = -10 kcal/mol). As a result of these technical limits, unfolded proteins are somewhat arbitrarily defined as having a $\Delta G_{unfolding}$ of less than -2 kcal/mol. Nevertheless, the 100,000-fold difference in binding affinity that can result from subtle differences in the relative stabilities of two "unfolded" molecules is enormous and likely to have important biological consequences.

The occasional appearance of natively disordered regions in protein structures has encouraged researchers to undertake a more systematic search for disordered regions in genome and sequence databases. Researchers reason that if local amino acid composition can be used to predict classical structural components such as alpha and beta folds (Anfinsen 1973), the same sequence data could also be used to anticipate the lack of secondary protein structure. This rational idea generated a series of predictors for identifying disordered regions among various data sets based on local factors such as amino acid composition, flexibility, hydrophathy, charge, and others (Romero, Obradovic et al. 1997; Romero P 1997; Dunker, Garner et al. 1998; Li, Obradovic et al. 2000; Dunker, Lawson et al. 2001)

By searching for amino acid sequences generally associated with naturally disordered regions, bioinformaticians have been able to show that a substantial fraction of sequenced proteins and coding regions are likely to have extensive disordered regions. The exact numbers vary based on the approach. However, studies have consistently detected significantly higher fractions of disordered regions in eukaryotic proteins compared to prokaryotes. Romero et al. reported in 1998 that a quarter of the proteins in the Swiss Prot database contained disordered regions in excess of 40 amino acids (Romero, Obradovic et al. 1998). Dunker et al.'s 2000 analysis of disordered regions in complete eukaryotic and prokaryotic genomes revealed that nearly a quarter of all eukaryotic proteins contain disordered regions exceeding 55 amino acids (Dunker, Obradovic et al. 2000). Using a similar analysis of the Protein Data Bank in 2002, Liu et al. observed that 20% of the eukaryotic and 4% of the prokaryotic proteins contained disordered regions of 70 or more amino acids (Liu, Tan et al. 2002). These initial studies have made it clear that disordered regions are as much a part of the protein landscape as the classically defined alpha helices, beta strands, and loops.

Despite their Lack of Structure, Disordered Regions Have Distinct Biological Function

As part of their 2002 analysis of the Protein Data Bank, Liu et al. noted that the disordered regions observed in eukaryotic and prokaryotic proteins appeared to be evolutionally conserved (Liu, Tan et al. 2002), suggesting that distinct biological functions and corresponding selective pressures are associated with disordered regions. That same year Dunker et al. published a literature review of the biological functions associated with 115 eukaryotic proteins containing predicted or observed disordered regions (Dunker, Brown et al. 2002). These functions included protein-protein binding, protein-nucleic acid interactions, phosphorylation, metal binding, and polymerization, just to name a few.

Most prominent among the roles identified by Dunker and his colleagues was that of protein-protein interactions, which described 54 of the 115 proteins studied (Dunker, Brown et al. 2002; Iakoucheva, Brown et al. 2002). In many cases the disordered regions were directly implicated in the cell-signaling interactions. A similar link between disordered regions and cell signaling had been made by Wright and Dyson in 1999 (Wright and Dyson 1999). On more than one occasion, researchers have speculated that the elevated presence of disordered regions in eukaryotes, when compared to similar structures in prokaryotes, is directly related to the need for signaling and regulatory components of more complex eukaryotic organisms (Dunker, Obradovic et al. 2000; Dunker, Lawson et al. 2001; Liu and Rost 2001; Iakoucheva, Brown et al. 2002; Vucetic, Brown et al. 2003).

There appear to be a number of reasons why disordered regions make excellent signal proteins. Chief among these are their abilities to bind multiple targets (Kriwacki, Hengst et al. 1996; Dyson and Wright 2002) with high specificity and low affinity (Schulz 1979; Dunker, Garner et al. 1998; Dunker, Lawson et al. 2001). Studies suggest that in most cases disordered regions bind their cognate ligands by adopting the fold necessary to bring the required residues in contact with binding epitope (Bourhis, Johansson et al. 2004; Callaghan, Aurikko et al. 2004; Lacy, Filippov et al. 2004; Oldfield, Cheng et al. 2005). Researchers have coined the term molecular recognition element (MoRE) to describe short disordered stretches of protein that form alpha, beta, irregular, or compound structures upon binding to their cognate ligand (CJ 2002; Bourhis, Johansson et al. 2004; Oldfield, Cheng et al. 2005; Uversky, Oldfield et al. 2005). Oldfield et al. provide a detailed review of 14 alpha-MoREs (Oldfield, Cheng et al. 2005), disordered regions that form alpha helices upon binding their cognate ligands. A more thorough analysis of the Protein Databank has revealed 2500 MoREs consisting of 372 non-redundant, heterologous structures that form alpha, beta, and irregular structures when complexed to ligands (Dunker, Cortese et al. 2005). The 4E binding protein 1 (4E-BP1) (Mader, Lee et al. 1995; Fletcher, McGuire et al. 1998; Fletcher and Wagner 1998), p54 tumor suppressor protein (Kussie, Gorina et al. 1996; Rustandi, Baldisseri et al. 2000), and p27 regulatory protein (Lacy, Filippov et al. 2004) provide examples of molecules containing disordered regions that adopt alpha and alpha-beta structures upon binding to ligands.

Because disordered regions lack intrinsic structure, they appear to be capable of adjusting their conformation to interface with a broad array of ligands (Young, Kirshenbaum et al. 1999; Kuznetsov and Rackovsky 2003). This means that a single MoRE can have the structural flexibility necessary to bind multiple ligands. Calmodulin (Ikura, Clore et al. 1992; Meador, Means et al. 1993), $p21^{Waf1/Cip1/Sidi1}$ (Kriwacki, Hengst et al. 1996), and a 54-kDa signal-recognition particle (SRP54) provide three examples of proteins that bind multiple ligands by means of disordered regions. In a 1991 review of flexible binding epitopes, Gellman suggests that methionine residues give calmodulin and a SRP54 the flexibility required to conform to multiple hydrophobic ligand-binding pockets (Gellman 1991).

The energy required to fold the unstructured region before binding may offer additional benefits for disordered regions involved in signaling pathways. Because unstructured proteins are required to adopt an energetically unfavorable conformation before binding their cognate ligands, the binding reaction exhibits reduced affinity with high specificity (Dunker, Garner et al. 1998). Dunker et al. argue that disordered regions are able to effectively form highly connected hubs in scale-free signaling networks precisely because they can bind multiple ligands with high specificity (Dunker, Cortese et al. 2005). The low binding affinity observed by disordered regions has the added advantage of limiting the signal duration so it can be quickly shut down.

It is worth noting that, despite the lower incidence of disordered regions in non-eukaryotic proteins, unstructured elements have also been shown to support membrane transport and macromolecular assembly in bacterial and viral proteins. For example, the HA2 domain of the influenza virus undergoes a conformational switch at low pH to support membrane insertion (Skehel and Wiley 2000). The bacterial FlgM flagella protein (Daughdrill, Chadsey et al. 1997; Samatey, Imada et al. 2001) and *Yersinia pestis* YopD effector protein (Raab and Swietnicki 2008) provide two examples of how disordered regions may support transport across bacterial membranes. The tobacco mosaic virus and bacterial flagellum also exemplify the role of disordered regions in macromolecule assembly, as reviewed by Namba (Namba 2001).

The *Bacillus subtilis* protease, subtilisin, provides another well-documented example of functional disorder in a prokaryotic protein. Subtilisin is activated when it auto cleaves its prodomain. In complex with subtilisin, the prodomain folds into a stable compact structure comprising a four-stranded antiparallel β-sheet and two three-turn α-helices. When isolated from subtilisin, the prodomain is 97% unfolded even under optimal folding conditions (Bryan, Wang et al. 1995). This low stability likely ensures that the prodomain-subtilisin complex exists only transiently and that the mature protease is released expeditiously in active form. Therefore, despite their lower incidence, the role of disordered regions in microbes is not to be entirely discounted.

Regardless of whether disordered regions are involved in cell signaling, membrane transport, or macromolecular assemble, or other cellular functions, it is clear from the proceeding examples that structure is not a necessary prerequisite for function. Rather, disordered regions appear to exhibit distinct functional advantages over highly structured elements. In the case of unstructured protein elements, binding drives folding rather than the other way around.

Ambiguities in the Protein Code Make Protein Structure Susceptible to a Range of Environmental Influences

Researchers have traditionally focused on identifying sequence patterns that have a propensity to adopt classic secondary structures such as alpha helices and beta strands (Rooman, Kocher et al. 1992; Bystroff, Simons et al. 1996; Bystroff, Thorsson et al. 2000; Jonassen, Eidhammer et al. 2000). And yet, despite the availability of rapidly expanding structural databases, the algorithms involved in sequence-based structural predictions had not exceeded a 60-70% accuracy rate as of 1993 (Gamier and Levin 1991; Rackovsky 1993). The limited accuracy of sequence-based structural predictions suggests an inherent ambiguity in protein code, which is supported by studies of chameleon sequences that are capable of adopting different structures based on distal or external structural cues. Multiple studies indicate that small stretches of amino acid sequences are capable of accommodating different secondary and tertiary structures (Minor and Kim 1996; Zhou, Alber et al. 2000; Kuznetsov and Rackovsky 2003) regardless of their amino acid composition. Independent analyses of structural databases reveal that sequences up to 9 amino acids occupy different tertiary structures (Kabsch and Sander 1983; Kabsch and Sander 1984; Wilson, Haft et al. 1985; Argos 1987; Cohen, Presnell et al. 1993; Mezei 1998; Sudarsanam 1998; Zhou, Alber et al. 2000; Kuznetsov and Rackovsky 2003). Zhou et al. specifically describe one 9-mer, which forms an alpha helix in one protein and a beta strand in another (Zhou, Alber et al. 2000). The same group noted that the helical peptide EAAAA occupies a beta strand in a hydrolase inhibitor (1HLRA) while the peptide VVVIV, which contains residues with a strong propensity for beta strands, adopts a helical structure in the lipoprotein 1SPF (Zhou, Alber et al. 2000). Minor and Kim showed how an 11-residue sequence adopted different secondary structures when placed in different positions of protein G (Minor and Kim 1996).

These discoveries are further supported by the efforts of various groups to engineer highly homologous protein pairs, which adopt different folds (protein heteromorphs) or create proteins that are capable of switching fold based on changes in environmental conditions. Both approaches offer further examples of the ambiguities inherent in the protein code. Responding to the 1994 Paracelsus challenge, which encouraged researchers to engineer two heteromorphic proteins with greater than 50% sequence identity (Rose and Creamer 1994), two groups independently created protein pairs, which adapted alpha or alpha-beta structures while exhibiting the required 50% sequence identity (Jones, Moody et al. 1996; Dalal, Balasubramanian et al. 1997). Dalal and Regan eventually constructed 60-residue heteromorphs with 66% and 86% identity (Dalal and Regan 2000). Our own lab at the University of Maryland's Center for Research and Biotechnology, was able to create two heteromorphic proteins with 88% sequence identity, thereby encoding two distinct 56 amino acid structures in only seven polymorphisms (Alexander, He et al. 2007). This was accomplished by embedding the functional epitopes for one structural species into the sequence of another. Phage display was used to select proteins that encoded the desired structure by panning for mutants whose conformation uncovered the cryptic binding epitope.

One of the more striking examples of two proteins with minimally encoded secondary structure arises from the work of Sauer's lab on the Arc repressor. The group found that simply swapping the positions of two adjacent asparagine and leucine residues transformed an antiparallel beta-sheet found in the Arc repressor homodimer to a pair of alpha helices (Cordes, Walsh et al. 1999). Sauer's lab subsequently reduced this structural switch to the single asparagine to leucine mutation, which drove 70% of the protein population to adapt a helical structure. The structural transformation was reversible and heavily dependent on temperature, solvent conditions, and ligand binding (Cordes, Burton et al. 2000).

In addition to demonstrating sequence ambiguity with highly homologous heteromorphs, scientists have also succeeded in engineering proteins that adopt different folds based on a variety of environmental cues such as temperature, pH, metal chelation, and phosphorylation. For example, several independent groups have designed proteins to switch folds when changes in pH alter the polarity of residues (Mutter, Gassmann et al. 1991; Cerpa, Cohen et al. 1996; Altman, Lee et al. 2000; Fezoui, Hartley et al. 2000; Li, Nicol et al. 2004; Lee and Chmielewski 2005). Kammerer et al. constructed a peptide that switches from an alpha coiled-coil structure to amyloid fibrils at elevated temperatures (Kammerer, Kostrewa et al. 2004) while another team created a high-temperature alpha to beta switch by simply destabilizing the coiled-coil structure of one protein (Ciani, Hutchinson et al. 2002). Patel et al. report that introducing histidines in a coiled-coil structure switches the protein to a beta-sheet in the presence of metal ions (Pagel, Vagt et al. 2005). Finally, several labs have reported the design of structural switches that respond to phosphorylation (Signarvic and DeGrado 2003; Balakrishnan and Zondlo 2006).

These man-made constructs are not without their natural analogs. The peer-reviewed literature contains occasional examples of well-ordered proteins that undergo profound conformational switches in response to their environment. For example, the hemaglutinin protein of the influenza virus is triggered by the low pH of intracellular compartments to adopt an elongated coiled-coil conformation, which facilitates membrane insertion (Skehel and Wiley 2000). G proteins adopt alternate conformations to modulate their interactions with different ligands depending on whether they are bound to GTP or GDP (Vetter and Wittinghofer 2001). The conformational switches in the prion protein are reportedly affected by changes in pH (Swietnicki, Petersen et al. 1997).

As with disordered regions, ordered regions appear to be influenced to adopt different folds when binding ligands. Ligands arguably exert a similar influence on local secondary structures as do more distal regions on the same protein, the only difference being that the latter is covalently linked to the region in question. Both protein folding and ligand binding offer similar—potentially competing—opportunities to bury hydrophobic residues. Therefore, it is not surprising that many examples exist for man-made and natural ligand-induced conformational switches (Stouten, Sander et al. 1993; Ashkenazi, Ripoll et al. 1997; Noel 1997; Solano, Fuertes et al. 1997; Azuma, Renault et al. 1999; Simeonidis, Stauber et al. 1999; Sola, Lopez-Hernandez et al. 2000; Falke, Fisher et al. 2001). Oligomerization is another case of ligand-induced conformational switching of which prions and other plaque-forming proteins provide an excellent example (Carrell and Gooptu 1998; Swietnicki, Morillas et al. 2000; Knaus, Morillas et al. 2001).

Environmental Factors Play a Major Role in Determining Protein Structure and Function

The thermodynamic model for protein folding teaches that proteins are dynamic molecules, which are in a state of flux. At equilibrium, it is possible to find populations of both folded and unfolded proteins, although the latter may be too small to measure. It is also clear from thermodynamic analysis that environmental factors such as temperature, pH, and protein concentrations can drive the equilibrium to favor the folded or unfolded states. This understanding is consistent with the research described above, which suggests that both disordered and ordered regions of proteins are capable of undergoing structural transformations based on an array of different environmental signals and cues. Nevertheless, mounting evidence for protein plasticity remains at odds with the classic notion of biology, which states that protein sequence specifies a unique fold which in turn determines function.

The prevalence of disordered regions in protein databases coupled with real-life examples of protein elements that fold upon ligand binding (Spolar and Record 1994; Demchenko 2001; Dyson and Wright 2002) have led many to question the classic view that structure is a prerequisite for function. It is increasingly clear that proteins don't need to exist as preformed structures before ligand binding takes place. In fact, steric factors occasionally prevent proteins from folding before binding. Calmodulin (Meador, Means et al. 1992) and TFIIIA (Choo and Schwabe 1998) are two examples of molecules for which folding and ligand binding are concurrent processes.

The accumulated evidence for functional disordered regions prompted Dunker and his colleagues to propose the trinity model for protein folding (Dunker, Lawson et al. 2001). As diagramed in equation 1, Dunker's model suggests that each protein has the potential to exist in one of three interconnected states: folded (F), globular (G), or unfolded (U), depending on environmental conditions.

$$F \Leftrightarrow G \Leftrightarrow U \Leftrightarrow F \tag{1}$$

Transitions between each of the states are brought about by ligand binding, phosphorylation, and other factors described above. Nor is protein function associated exclusively with its folded state, as it is occasionally necessary for a protein to exist in a molten or unfolded form to transition membranes or interface with ligands. In presenting their model, Dunker and his coauthors set out a number of examples in which transitions between folded, globular, and unfolded states are necessary for protein function.

The trinity model was proposed to account for the diverse array of function observed in natively disordered regions. However, the model does not explicitly account for the potential of folded, globular, and unstructured proteins to adopt more than one fold. Equation 2 expands upon the trinity model by allowing for a transition between alternate folds (F_A or F_B) via an unfolded intermediate (U). Each of the folded conformations facilitates and is potentially induced by an interaction with a corresponding ligand (L_A or L_B). While only two conformations are depicted in Equation 2, others are possible. Furthermore, the difference between conformations need not be as drastic as a shift between alpha and beta folds, although this is possible. Rather F_A and F_B could represent the two different alpha-MoRE conformations required to bind different ligands.

$$F_A \bullet L_A \Leftrightarrow F_A + L_A \Leftrightarrow U \Leftrightarrow F_B + L_B \Leftrightarrow F_B \bullet L_B \tag{2}$$

As with any thermodynamic equation, environmental parameters such as protein and ligand concentrations, temperature, and pH can skew the steady state equilibrium in favor of one or another of the states. Genetic factors can also affect this balance. Limited largely by steric conflicts, a polypeptide chain can potentially adopt a broad range of conformations. However, all but a few of these conformational states are thermodynamically unfavorable and are therefore minimally occupied in the steady state. Nevertheless, it is easy to imagine that a single core mutation could sufficiently destabilize F_A in favor of U or F_B. Conversely, F_A stabilizing mutations could effectively exclude the protein from any of the other states. In this way it is possible to account for the functional plasticity observed in a growing number of proteins without excluding the large number of classically structured proteins. More importantly, this analysis underscores a critical role for amino acid sequences and sequence polymorphisms in regulating the range of potential peptide conformations and interactions rather than simply specifying absolute folds and functions.

Revisiting Biology's Central Dogma and Considering the Role of Disordered Regions in Protein Evolution

As mentioned at the outset, the central dogma of biology describes a direct correlation between genotype and phenotype in which the link between protein sequence, structure, and function plays a critical role. Despite efforts to unambiguously tie polypeptide sequences to protein structures, a fair degree of ambiguity remains. This is evidenced by the systematic identification of functionally disordered regions in 25-30% of eukaryotic proteins (Oldfield, Cheng et al. 2005) and the identification of structurally ambivalent sequences in protein databases. In both cases, it is apparent that environmental factors play important roles in determining protein structure.

It is therefore worth considering whether a relaxed interpretation of the central dogma may offer a more accurate and informative view of protein biology. Rather than requiring one-to-one correlations between protein sequences and structures as implied by the current paradigm, it may be beneficial to explicitly allow for one-to-many relationships in which genetically encoded protein sequences potentially define a broad range of loosely folded alpha or beta structures that are further specified by environmental factors such as ligand concentrations, pH, and post-translational modifications. Many sequences will rigidly define the protein structure and function in the classical sense. However, others will allow a great deal of flexibility to the extent that protein regions are disordered under many physiological conditions. Thus, the pathway from genotype to phenotype becomes a more loosely defined process with important implications for our understanding of protein evolution.

According to a one-to-many interpretation of the central dogma, single point mutations are capable of dramatically altering the protein's conformational stability and the extent to which environmental conditions, including protein-ligand interactions, can impact the molecule's structure and function. As we have already seen, protein segments existing in molten or disordered states are capable of exploring a broader array of conformations, which can be catalyzed by novel binding interactions. Favorable functions would be expected to drive selected enhancements to protein affinity and specificity through point mutations that stabilize the protein core or improve contacts in the binding epitope. In this manner, structural ambiguities of the types engendered in the protein code, can be seen as a prerequisites for the evolution of novel function.

Our own observation that a latent binding epitope can be engineered into a small protein without interfering with its native function (Alexander, He et al. 2007) suggests that the "folding problem" might be viewed from a new perspective. The vast array of conformations available to a polypeptide chain means that many latent functions could exist in this vast population of conformers. Because of the prevailing view that all folds except the native are highly improbable, the acquisition of new function by conformational switching also is judged to be improbable. Because the conformational preference can depend on a delicate balance of critical interactions within a protein, however, a few mutations can result in a new conformation and the unmasking of new functionality.

Numerous examples exist of natural proteins which are largely unfolded unless bound to a ligand. We would suggest that transitional forms in the natural evolution of new folds may be predominantly unfolded (Ward, Sodhi et al. 2004). Because 25-30% of the eukaryotic proteome is predicted to be "natively unfolded," innovation (i.e., unmasking of new function) could result from a few mutations that could lower the propensity for the native fold but increase the propensity for an alternative one.

The propensity for disordered regions to uncover novel function is nicely illustrated in a study by Brown et al., who observed increased rates of evolutionary divergence among disordered regions in yeast-encoded proteins (Brown, Takayama et al. 2002). According to the authors of this study, 19 of 26 protein families studied showed increased evolutionary change in their disordered regions, five showed no difference between ordered and disordered regions, and one exhibited the reverse phenomena. Others have observed similarly increased evolutionary diversity among the disordered regions of individual proteins such as topoisomerase II (Shaiu, Hu et al. 1999), ribosomal protein S4 (Sayers, Gerstner et al. 2000), and potassium channel subunits (Wissmann, Baukrowitz et al. 1999). Another report suggests increased promiscuity among disordered regions by observing more elevated disordered regions among cancer-associated signaling proteins than those of healthy cells (Iakoucheva, Brown et al. 2002).

The hypothesis that global or local protein disorder represents a valid transition state in the evolution of novel biological function requires a genetically defined pathway by which one highly stabilized fold can transition to another. A genetic mechanism for individually regulating the structural stability and ligand binding propensity of proteins has already been postulated by Dunker et al. who suggested that affinity and specificity can be independently modulated by mutations to protein core and ligand interface residues, respectively (Dunker, Garner et al. 1998). The group's model, which was inspired by Schulz (Schulz 1977), describes how core mutations can relax protein structure by disrupting hydrophilic packing. Dunker's team suggests that destabilizing core mutations make it possible for disordered regions to decouple affinity and specificity in a manner that is not possible for highly structured proteins. According to the group's analysis, the free energy consumed by adopting the induced-fold required for ligand binding reduces the overall energetics of the reaction while simultaneously increasing the ratio of disassociation and association rates for the interaction (i.e., binding specificity). Beyond accounting for the independent regulation of affinity and specificity, the group's model suggests a mechanism by which destabilizing core mutations thermodynamically free proteins to explore and refine alternate binding interfaces. Conversely, mutations that stabilize the core of ligand-bound proteins limit the ability of the molecule to interact with other partners and stabilize a biologically beneficial function. The latter process is consistent with phage display experiments in which enhanced ligand binding is achieved through core-stabilizing mutations that preserve the active conformation (Gu, Yi et al. 1995; Ruan, Hoskins et al. 1998; Rozak, Alexander et al. 2006).

While the proteins involved in prion encephalopathies, such as bovine spongiform encephalopathy, Creutzfeldt-Jakob disease, and Alzheimer's disease represent maladapted structures that induce the formation of disease-causing plaques, we propose that these proteins may provide natural examples of the evolutionary drift between diverse structures and function. Inheritable forms of prion encephalopathy have been associated with more than 20 distinct mutations in the PrP prion protein gene (Hsiao, Baker et al. 1989). Each of these mutations appear to make the plaque-forming PrP^{SC} variant thermodynamically accessible by destabilizing the structural core of the healthy PrP^{C} isoform (Huang, Gabriel et al. 1994; Riek, Hornemann et al. 1996; Prusiner, Scott et al. 1998). The prion protein is just one of many proteins that cause disease by adopting amyloid plaque and fibril-forming isoforms. Serpin protease inhibitors, haemoglobin, and the immunoglobulin light chain are just a few of the proteins that have been implicated in conformational diseases (Carrell and Gooptu 1998). More than 50 distinct mutations have been attributed to disease-causing conformational switches in serpins alone (Stein and Carrell 1995; Carrell, Lomas et al. 1997). Prion-like proteins have also been identified and studied in yeast (Wickner 1994; Chernoff, Lindquist et al. 1995; Paushkin, Kushnirov et al. 1996) and fungi (Coustou, Deleu et al. 1997; Wickner 1997; Coustou, Deleu et al. 1999). Each of these examples attest to the widespread effect of destabilizing mutation on cell health and may point to the underlying role of similar mutations in protein evolution.

CONCLUSION

Proteins occasionally exhibit the potential to adopt alternate folds under natural or artificial conditions. These conformational changes, which are frequently induced by oligomerization, can lead to the formation of disease-forming amyloid plaques or accumulation of insoluble aggregates in over-expressed recombinant proteins. Because alternate folds challenge traditional notions of well-behaved proteins and frequently lead to problems in natural and laboratory settings, it is tempting to characterize these proteins as being misfolded or misbehaved. Unfortunately, this perspective masks the positive role structural plasticity can play in protein function and evolution. Indeed, the literature offers multiple examples of proteins that have either evolved or been engineered to solve natural and man-made problems by dramatically altering their secondary structures in response to physiological conditions. With an estimated 30% of eukaryotic proteins containing extensive unstructured regions and mutagenesis experiments showing that fewer than 12 amino acids in 100 are capable of specifying dramatically different structures, disordered regions and conformational switches may play a greater role in cell biology than previously expected. We have reviewed the expansive body of published research that supports a model for protein evolution in which conformational switching plays key roles in exposing cryptic epitopes and uncovering novel functions. Under this model, conformational diseases represent a negative outcome of a natural process that helps organisms explore available sequence space for positive biological activity.

REFERENCES

Alexander, P. A., Y. He, et al. (2007). "The design and characterization of two proteins with 88% sequence identity but different structure and function." *Proc Natl Acad Sci U S A 104*(29): 11963-8.

Altman, M., P. Lee, et al. (2000). "Conformational behavior of ionic self-complementary peptides." *Protein Sci 9*(6): 1095-105.

Anfinsen, C. B. (1973). "Principles that govern the folding of protein chains." *Science 181*(96): 223-30.

Anfinsen, C. B., E. Haber, et al. (1961). "The kinetics of formation of native ribonuclease during oxidation of the reduced polypeptide chain." *Proc Natl Acad Sci U S A 47*: 1309-14.

Argos, P. (1987). "Analysis of sequence-similar pentapeptides in unrelated protein tertiary structures. Strategies for protein folding and a guide for site-directed mutagenesis." *J Mol Biol 197*(2): 331-48.

Ashkenazi, G., D. R. Ripoll, et al. (1997). "A molecular switch for biochemical logic gates: conformational studies." *Biosens Bioelectron 12*(2): 85-95.

Azuma, Y., L. Renault, et al. (1999). "Model of the ran-RCC1 interaction using biochemical and docking experiments." *J Mol Biol 289*(4): 1119-30.

Balakrishnan, S. and N. J. Zondlo (2006). "Design of a protein kinase-inducible domain." *J Am Chem Soc 128*(17): 5590-1.

Biou, V., J. F. Gibrat, et al. (1988). "Secondary structure prediction: combination of three different methods." *Protein Eng 2*(3): 185-91.

Bohr, H., J. Bohr, et al. (1988). "Protein secondary structure and homology by neural networks. The alpha-helices in rhodopsin." *FEBS Lett 241*(1-2): 223-8.

Bourhis, J. M., K. Johansson, et al. (2004). "The C-terminal domain of measles virus nucleoprotein belongs to the class of intrinsically disordered proteins that fold upon binding to their physiological partner." *Virus Res 99*(2): 157-67.

Brown, C. J., S. Takayama, et al. (2002). "Evolutionary rate heterogeneity in proteins with long disordered regions." *J Mol Evol 55*(1): 104-10.

Bryan, P., L. Wang, et al. (1995). "Catalysis of a protein folding reaction: mechanistic implications of the 2.0 A structure of the subtilisin-prodomain complex." *Biochemistry 34*(32): 10310-8.

Bystroff, C., K. T. Simons, et al. (1996). "Local sequence-structure correlations in proteins." *Curr Opin Biotechnol 7*(4): 417-21.

Bystroff, C., V. Thorsson, et al. (2000). "HMMSTR: a hidden Markov model for local sequence-structure correlations in proteins." *J Mol Biol 301*(1): 173-90.

Callaghan, A. J., J. P. Aurikko, et al. (2004). "Studies of the RNA degradosome-organizing domain of the Escherichia coli ribonuclease RNase E." *J Mol Biol 340*(5): 965-79.

Carrell, R., D. Lomas, et al. (1997). "Dysfunctional variants and the structural biology of the serpins." *Adv Exp Med Biol 425*: 207-22.

Carrell, R. W. and B. Gooptu (1998). "Conformational changes and disease--serpins, prions and Alzheimer's." *Curr Opin Struct Biol 8*(6): 799-809.

Cerpa, R., F. E. Cohen, et al. (1996). "Conformational switching in designed peptides: the helix/sheet transition." *Fold Des 1*(2): 91-101.

Chernoff, Y. O., S. L. Lindquist, et al. (1995). "Role of the chaperone protein Hsp104 in propagation of the yeast prion-like factor [psi+]." *Science 268*(5212): 880-4.

Choo, Y. and J. W. Schwabe (1998). "All wrapped up." *Nat Struct Biol 5*(4): 253-5.

Chou, P. Y. and G. D. Fasman (1978). "Prediction of the secondary structure of proteins from their amino acid sequence." *Adv Enzymol Relat Areas Mol Biol 47*: 45-148.

Ciani, B., E. G. Hutchinson, et al. (2002). "A designed system for assessing how sequence affects alpha to beta conformational transitions in proteins." *J Biol Chem 277*(12): 10150-5.

CJ, O. (2002). Predictions of disordered proteins and protein binding regions: applications to structural genomics. Washington, Washington State University Pullman.

Cohen, B. I., S. R. Presnell, et al. (1993). "Origins of structural diversity within sequentially identical hexapeptides." *Protein Sci 2*(12): 2134-45.

Cordes, M. H., R. E. Burton, et al. (2000). "An evolutionary bridge to a new protein fold." *Nat Struct Biol 7*(12): 1129-32.

Cordes, M. H., N. P. Walsh, et al. (1999). "Evolution of a protein fold in vitro." *Science 284*(5412): 325-8.

Coustou, V., C. Deleu, et al. (1997). "The protein product of the het-s heterokaryon incompatibility gene of the fungus Podospora anserina behaves as a prion analog." *Proc Natl Acad Sci U S A 94*(18): 9773-8.

Coustou, V., C. Deleu, et al. (1999). "Mutational analysis of the [Het-s] prion analog of Podospora anserina. A short N-terminal peptide allows prion propagation." *Genetics 153*(4): 1629-40.

Dalal, S., S. Balasubramanian, et al. (1997). "Protein alchemy: changing beta-sheet into alpha-helix." *Nat Struct Biol 4*(7): 548-52.

Dalal, S. and L. Regan (2000). "Understanding the sequence determinants of conformational switching using protein design." *Protein Sci 9*(9): 1651-9.

Daughdrill, G. W., M. S. Chadsey, et al. (1997). "The C-terminal half of the anti-sigma factor, FlgM, becomes structured when bound to its target, sigma 28." *Nat Struct Biol 4*(4): 285-91.

Demchenko, A. P. (2001). "Recognition between flexible protein molecules: induced and assisted folding." *J Mol Recognit 14*(1): 42-61.

Dunker, A. K., C. J. Brown, et al. (2002). "Intrinsic disorder and protein function." *Biochemistry 41*(21): 6573-82.

Dunker, A. K., M. S. Cortese, et al. (2005). "Flexible nets. The roles of intrinsic disorder in protein interaction networks." *Febs J 272*(20): 5129-48.

Dunker, A. K., E. Garner, et al. (1998). "Protein disorder and the evolution of molecular recognition: theory, predictions and observations." *Pac Symp Biocomput*: 473-84.

Dunker, A. K., J. D. Lawson, et al. (2001). "Intrinsically disordered protein." *J Mol Graph Model 19*(1): 26-59.

Dunker, A. K., Z. Obradovic, et al. (2000). "Intrinsic protein disorder in complete genomes." *Genome Inform Ser Workshop Genome Inform 11*: 161-71.

Dyson, H. J. and P. E. Wright (2002). "Coupling of folding and binding for unstructured proteins." *Curr Opin Struct Biol 12*(1): 54-60.

Falke, S., M. T. Fisher, et al. (2001). "Structural changes in GroEL effected by binding a denatured protein substrate." *J Mol Biol 308*(4): 569-77.

Fezoui, Y., D. M. Hartley, et al. (2000). "A de novo designed helix-turn-helix peptide forms nontoxic amyloid fibrils." *Nat Struct Biol 7*(12): 1095-9.

Fletcher, C. M., A. M. McGuire, et al. (1998). "4E binding proteins inhibit the translation factor eIF4E without folded structure." *Biochemistry 37*(1): 9-15.

Fletcher, C. M. and G. Wagner (1998). "The interaction of eIF4E with 4E-BP1 is an induced fit to a completely disordered protein." *Protein Sci 7*(7): 1639-42.

Gamier, J. and J. M. Levin (1991). "The protein structure code: what is its present. status? ." *Comput. Applic. Biosci. 7*: 133-142.

Gellman, S. H. (1991). "On the role of methionine residues in the sequence-independent recognition of nonpolar protein surfaces." *Biochemistry 30*(27): 6633-6.

Gu, H., Q. Yi, et al. (1995). "A phage display system for studying the sequence determinants of protein folding." *Protein Sci 4*(6): 1108-17.

Holley, L. H. and M. Karplus (1989). "Protein secondary structure prediction with a neural network." *Proc Natl Acad Sci U S A 86*(1): 152-6.

Hsiao, K., H. F. Baker, et al. (1989). "Linkage of a prion protein missense variant to Gerstmann-Straussler syndrome." *Nature 338*(6213): 342-5.

Huang, Z., J. M. Gabriel, et al. (1994). "Proposed three-dimensional structure for the cellular prion protein." *Proc Natl Acad Sci U S A 91*(15): 7139-43.

Huntley, M. A. and G. B. Golding (2002). "Simple sequences are rare in the Protein Data Bank." *Proteins 48*(1): 134-40.

Iakoucheva, L. M., C. J. Brown, et al. (2002). "Intrinsic disorder in cell-signaling and cancer-associated proteins." *J Mol Biol 323*(3): 573-84.

Ikura, M., G. M. Clore, et al. (1992). "Solution structure of a calmodulin-target peptide complex by multidimensional NMR." *Science 256*(5057): 632-8.

Jonassen, I., I. Eidhammer, et al. (2000). "Searching the protein structure databank with weak sequence patterns and structural constraints." *J Mol Biol 304*(4): 599-619.

Jones, D. T., C. M. Moody, et al. (1996). "Towards meeting the Paracelsus Challenge: The design, synthesis, and characterization of paracelsin-43, an alpha-helical protein with over 50% sequence identity to an all-beta protein." *Proteins 24*(4): 502-13.

Kabsch, W. and C. Sander (1983). "Dictionary of protein secondary structure: pattern recognition of hydrogen-bonded and geometrical features." *Biopolymers 22*(12): 2577-637.

Kabsch, W. and C. Sander (1984). "On the use of sequence homologies to predict protein structure: identical pentapeptides can have completely different conformations." *Proc Natl Acad Sci U S A 81*(4): 1075-8.

Kammerer, R. A., D. Kostrewa, et al. (2004). "Exploring amyloid formation by a de novo design." *Proc Natl Acad Sci U S A 101*(13): 4435-40.

Knaus, K. J., M. Morillas, et al. (2001). "Crystal structure of the human prion protein reveals a mechanism for oligomerization." *Nat Struct Biol 8*(9): 770-4.

Kriwacki, R. W., L. Hengst, et al. (1996). "Structural studies of p21Wafl/Cipl/Sdil in the free and Cdk2-bound state: conformational disorder mediates binding diversity." *Proc Natl Acad Sci U S A 93*(21): 11504-9.

Kussie, P. H., S. Gorina, et al. (1996). "Structure of the MDM2 oncoprotein bound to the p53 tumor suppressor transactivation domain." *Science 274*(5289): 948-53.

Kuznetsov, I. B. and S. Rackovsky (2003). "On the properties and sequence context of structurally ambivalent fragments in proteins." *Protein Sci 12*(11): 2420-33.

Lacy, E. R., I. Filippov, et al. (2004). "p27 binds cyclin-CDK complexes through a sequential mechanism involving binding-induced protein folding." *Nat Struct Mol Biol 11*(4): 358-64.

Lee, H. M. and J. Chmielewski (2005). "Liposomal cargo unloading induced by pH-sensitive peptides." *J Pept Res 65*(3): 355-63.

Levin, J. M., B. Robson, et al. (1986). "An algorithm for secondary structure determination in proteins based on sequence similarity." *FEBS Lett 205*(2): 303-8.

Li, W., F. Nicol, et al. (2004). "GALA: a designed synthetic pH-responsive amphipathic peptide with applications in drug and gene delivery." *Adv Drug Deliv Rev 56*(7): 967-85.

Li, X., Z. Obradovic, et al. (2000). "Comparing predictors of disordered protein." *Genome Inform Ser Workshop Genome Inform 11*: 172-84.

Liu, J. and B. Rost (2001). "Comparing function and structure between entire proteomes." *Protein Sci 10*(10): 1970-9.

Liu, J., H. Tan, et al. (2002). "Loopy proteins appear conserved in evolution." *J Mol Biol 322*(1): 53-64.

Mader, S., H. Lee, et al. (1995). "The translation initiation factor eIF-4E binds to a common motif shared by the translation factor eIF-4 gamma and the translational repressors 4E-binding proteins." *Mol Cell Biol 15*(9): 4990-7.

Meador, W. E., A. R. Means, et al. (1992). "Target enzyme recognition by calmodulin: 2.4 A structure of a calmodulin-peptide complex." *Science 257*(5074): 1251-5.

Meador, W. E., A. R. Means, et al. (1993). "Modulation of calmodulin plasticity in molecular recognition on the basis of x-ray structures." *Science 262*(5140): 1718-21.

Mezei, M. (1998). "Chameleon sequences in the PDB." *Protein Eng 11*(6): 411-4.

Minor, D. L., Jr. and P. S. Kim (1996). "Context-dependent secondary structure formation of a designed protein sequence." *Nature 380*(6576): 730-4.

Mutter, M., R. Gassmann, et al. (1991). "Switch peptides - pH-induced alpha-helix to beta-sheet transitions of bisamphiphilic oligopeptides." *Angew. Chem. Int. Ed. Engl 30*: 1514-1516.

Namba, K. (2001). "Roles of partly unfolded conformations in macromolecular self-assembly." *Genes Cells 6*(1): 1-12.

Nishikawa, K. and T. Ooi (1986). "Amino acid sequence homology applied to the prediction of protein secondary structures, and joint prediction with existing methods." *Biochim Biophys Acta 871*(1): 45-54.

Noel, J. P. (1997). "Turning off the Ras switch with the flick of a finger." *Nat Struct Biol 4*(9): 677-80.

Okada, K., K. Hirotsu, et al. (2001). "Structures of Escherichia coli branched-chain amino acid aminotransferase and its complexes with 4-methylvalerate and 2-methylleucine: induced fit and substrate recognition of the enzyme." *Biochemistry 40*(25): 7453-63.

Oldfield, C. J., Y. Cheng, et al. (2005). "Comparing and combining predictors of mostly disordered proteins." *Biochemistry 44*(6): 1989-2000.

Oldfield, C. J., Y. Cheng, et al. (2005). "Coupled folding and binding with alpha-helix-forming molecular recognition elements." *Biochemistry 44*(37): 12454-70.

Pagel, K., T. Vagt, et al. (2005). "From alpha-helix to beta-sheet--a reversible metal ion induced peptide secondary structure switch." *Org Biomol Chem 3*(14): 2500-2.

Pain, R. H. and B. Robson (1970). "Analysis of the code relating sequence to secondary structure in proteins." *Nature 227*(5253): 62-3.

Pan, K. M., M. Baldwin, et al. (1993). "Conversion of alpha-helices into beta-sheets features in the formation of the scrapie prion proteins." *Proc Natl Acad Sci U S A 90*(23): 10962-6.

Paushkin, S. V., V. V. Kushnirov, et al. (1996). "Propagation of the yeast prion-like [psi+] determinant is mediated by oligomerization of the SUP35-encoded polypeptide chain release factor." *Embo J 15*(12): 3127-34.

Prusiner, S. B. (1982). "Novel proteinaceous infectious particles cause scrapie." *Science 216*(4542): 136-44.

Prusiner, S. B., M. R. Scott, et al. (1998). "Prion protein biology." *Cell 93*(3): 337-48.

Prusiner, S. B., M. R. Scott, et al. (1998). "Prion Protein Biology." *Cell 93*(3): 337-348.

Qian, N. and T. J. Sejnowski (1988). "Predicting the secondary structure of globular proteins using neural network models." *J Mol Biol 202*(4): 865-84.

Raab, R. and W. Swietnicki (2008). "Yersinia pestis YopD 150-287 fragment is partially unfolded in the native state." *Protein Expr Purif 58*(1): 53-60.

Rackovsky, S. (1993). "On the nature of the protein folding code." *Proc Natl Acad Sci U S A 90*(2): 644-8.

Riek, R., S. Hornemann, et al. (1996). "NMR structure of the mouse prion protein domain PrP(121-321)." *Nature 382*(6587): 180-2.

Romero, Obradovic, et al. (1997). "Sequence Data Analysis for Long Disordered Regions Prediction in the Calcineurin Family." *Genome Inform Ser Workshop Genome Inform 8*: 110-124.

Romero, P., Z. Obradovic, et al. (1998). "Thousands of proteins likely to have long disordered regions." *Pac Symp Biocomput*: 437-48.

Romero P, O. Z., Kissinger C, Villafranca JE, Dunker AK (1997). *Identifying disordered regions in proteins from amino acid sequence*.

Rooman, M. J., J. P. Kocher, et al. (1992). "Extracting information on folding from the amino acid sequence: accurate predictions for protein regions with preferred conformation in the absence of tertiary interactions." *Biochemistry 31*(42): 10226-38.

Rose, G. D. and T. P. Creamer (1994). "Protein folding: predicting predicting." *Proteins 19*(1): 1-3.

Rozak, D. A., P. A. Alexander, et al. (2006). "Using offset recombinant polymerase chain reaction to identify functional determinants in a common family of bacterial albumin binding domains." *Biochemistry 45*(10): 3263-71.

Ruan, B., J. Hoskins, et al. (1998). "Stabilizing the subtilisin BPN' pro-domain by phage display selection: how restrictive is the amino acid code for maximum protein stability?" *Protein Sci 7*(11): 2345-53.

Rustandi, R. R., D. M. Baldisseri, et al. (2000). "Structure of the negative regulatory domain of p53 bound to S100B(betabeta)." *Nat Struct Biol 7*(7): 570-4.

Samatey, F. A., K. Imada, et al. (2001). "Structure of the bacterial flagellar protofilament and implications for a switch for supercoiling." *Nature 410*(6826): 331-7.

Sayers, E. W., R. B. Gerstner, et al. (2000). "Structural preordering in the N-terminal region of ribosomal protein S4 revealed by heteronuclear NMR spectroscopy." *Biochemistry 39*(44): 13602-13.

Schiffer, M. and A. B. Edmundson (1967). "Use of helical wheels to represent the structures of proteins and to identify segments with helical potential." *Biophys J 7*(2): 121-35.

Schulz, G. E. (1977). Nucleotide Binding Proteins. *Molecular Mechanisms of Biological Recognition*. M. Balaban. New York, Elsevier/North-Holland: 79-94.

Schulz, G. E. (1979). Nucleotide Binding Proteins. *Molecular Mechanisms of Biological Recognition*. M. Balaban. New York, Elsevier/North-Holland: 79-94.

Shaiu, W. L., T. Hu, et al. (1999). "The hydrophilic, protease-sensitive terminal domains of eucaryotic DNA topoisomerases have essential intracellular functions." *Pac Symp Biocomput*: 578-89.

Signarvic, R. S. and W. F. DeGrado (2003). "De novo design of a molecular switch: phosphorylation-dependent association of designed peptides." *J Mol Biol 334*(1): 1-12.

Simeonidis, S., D. Stauber, et al. (1999). "Mechanisms by which IkappaB proteins control NF-kappaB activity." *Proc Natl Acad Sci U S A 96*(1): 49-54.

Skehel, J. J. and D. C. Wiley (2000). "Receptor binding and membrane fusion in virus entry: the influenza hemagglutinin." *Annu Rev Biochem 69*: 531-69.

Sola, M., E. Lopez-Hernandez, et al. (2000). "Towards understanding a molecular switch mechanism: thermodynamic and crystallographic studies of the signal transduction protein CheY." *J Mol Biol 303*(2): 213-25.

Solano, R., A. Fuertes, et al. (1997). "A single residue substitution causes a switch from the dual DNA binding specificity of plant transcription factor MYB.Ph3 to the animal c-MYB specificity." *J Biol Chem 272*(5): 2889-95.

Spolar, R. S. and M. T. Record, Jr. (1994). "Coupling of local folding to site-specific binding of proteins to DNA." *Science 263*(5148): 777-84.

Stein, P. E. and R. W. Carrell (1995). "What do dysfunctional serpins tell us about molecular mobility and disease?" *Nat Struct Biol 2*(2): 96-113.

Stouten, P. F., C. Sander, et al. (1993). "How does the switch II region of G-domains work?" *FEBS Lett 320*(1): 1-6.

Sudarsanam, S. (1998). "Structural diversity of sequentially identical subsequences of proteins: identical octapeptides can have different conformations." *Proteins 30*(3): 228-31.

Sweet, R. M. (1986). "Evolutionary similarity among peptide segments is a basis for prediction of protein folding." *Biopolymers 25*(8): 1565-77.

Swietnicki, W., M. Morillas, et al. (2000). "Aggregation and fibrillization of the recombinant human prion protein huPrP90-231." *Biochemistry 39*(2): 424-31.

Swietnicki, W., R. Petersen, et al. (1997). "pH-dependent stability and conformation of the recombinant human prion protein PrP(90-231)." *J Biol Chem 272*(44): 27517-20.

Uversky, V. N., J. R. Gillespie, et al. (2000). "Why are "natively unfolded" proteins unstructured under physiologic conditions?" *Proteins 41*(3): 415-27.

Uversky, V. N., C. J. Oldfield, et al. (2005). "Showing your ID: intrinsic disorder as an ID for recognition, regulation and cell signaling." *J Mol Recognit 18*(5): 343-84.

Vetter, I. R. and A. Wittinghofer (2001). "The guanine nucleotide-binding switch in three dimensions." *Science 294*(5545): 1299-304.

Vucetic, S., C. J. Brown, et al. (2003). "Flavors of protein disorder." *Proteins 52*(4): 573-84.

Ward, J. J., J. S. Sodhi, et al. (2004). "Prediction and functional analysis of native disorder in proteins from the three kingdoms of life." *J Mol Biol 337*(3): 635-45.

Weiss, M. A. (2001). "Floppy SOX: mutual induced fit in hmg (high-mobility group) box-DNA recognition." *Mol Endocrinol 15*(3): 353-62.

Wickner, R. B. (1994). "[URE3] as an altered URE2 protein: evidence for a prion analog in Saccharomyces cerevisiae." *Science 264*(5158): 566-9.

Wickner, R. B. (1997). "A new prion controls fungal cell fusion incompatibility." *Proc Natl Acad Sci U S A 94*(19): 10012-4.

Wilson, I. A., D. H. Haft, et al. (1985). "Identical short peptide sequences in unrelated proteins can have different conformations: a testing ground for theories of immune recognition." *Proc Natl Acad Sci U S A 82*(16): 5255-9.

Wissmann, R., T. Baukrowitz, et al. (1999). "NMR structure and functional characteristics of the hydrophilic N terminus of the potassium channel beta-subunit Kvbeta1.1." *J Biol Chem 274*(50): 35521-5.

Wright, P. E. and H. J. Dyson (1999). "Intrinsically unstructured proteins: re-assessing the protein structure-function paradigm." *J Mol Biol 293*(2): 321-31.

Wyatt, R., P. D. Kwong, et al. (1998). "The antigenic structure of the HIV gp120 envelope glycoprotein." *Nature 393*(6686): 705-11.

Yaremchuk, A., M. Tukalo, et al. (2001). "A succession of substrate induced conformational changes ensures the amino acid specificity of Thermus thermophilus prolyl-tRNA synthetase: comparison with histidyl-tRNA synthetase." *J Mol Biol 309*(4): 989-1002.

Young, M., K. Kirshenbaum, et al. (1999). "Predicting conformational switches in proteins." *Protein Sci 8*(9): 1752-64.

Zhou, X., F. Alber, et al. (2000). "An analysis of the helix-to-strand transition between peptides with identical sequence." *Proteins 41*(2): 248-56.

In: Protein Misfolding
Editors: C. B. O'Doherty, A.C. Byrne
ISBN: 978-1-60456-881-3

Chapter III

PROTEIN TRANSLATION RATES AND PROTEIN MISFOLDING: IS THERE ANY LINK?

Anton A. Komar
Department of Biological, Geological and Environmental Sciences
Cleveland State University, Cleveland, OH 44115, USA

ABSTRACT

Protein folding remains one of the most puzzling problems in molecular biology. Protein folding is now widely assumed to be co-translational occurring as the polypeptide chain is being synthesized on the ribosome. With recognition of the co-translational nature of *in vivo* protein folding, the focus of many recent studies has shifted from the investigation of protein folding in a test tube (from the unfolded state) to the investigation of protein folding in the cell or in cell-free translation systems. Interest in the problem of *in vivo* protein folding has increased tremendously in recent years with the broad recognition of the central role of protein misfolding in many degenerative diseases and pathological conditions as well as with the increased need to produce correctly folded recombinant and therapeutic proteins. Numerous quality-control mechanisms exist in vivo which aid identification and removal of abnormally folded and misfolded proteins. Cells evolved two major quality control systems to deal with misfolded proteins: molecular chaperones which help to stabilize proteins and bring them back to the productive folding pathway, and the ubiquitin proteasome system, which degrades misfolded proteins that can not be refolded to the native state. The phenomenon of co-translational ubiquitination (and degradation) of proteins suggests that misfolding of proteins can start very early during protein synthesis on the ribosome. What are the factors influencing the ability of proteins to fold correctly during the process of translation? A number of experimental data and recent *in silico* modeling experiments suggest that the rate at which various portions of the growing polypeptide chain emerge from the ribosome can influence both the folding pathway and the final conformation of the protein. It has been thus suggested that abnormal translation rates may lead to increased levels of protein misfolding. This review summarizes recent developments and the prevalent literature covering this topic.

1. Introduction: Protein Folding and Misfolding — an Old and yet Unsolved Enigma

For decades the mechanism by which the polypeptide chain acquires its native structure has been investigated mainly by *in vitro* denaturation/renaturation experiments. These experiments provided a wealth of information and suggested that protein folding obeys a sequential model which postulates a unique pathway with defined intermediates [1]. Subsequently, the concept of the folding funnel has been developed suggesting that there could be multiple pathways which guide protein folding to a native conformation (state) characterized by the free energy minimum [2,3].

Many proteins can be successfully refolded in aqueous solutions, and these observations support the Anfinsen's principle which states that the information necessary to specify the native three-dimensional structure of a protein — which largely defines its function — is contained solely in its amino acid sequence [4]. Yet, many attempts to achieve *in vitro* 100% refolding of the isolated denatured proteins were only partially successful. This showed that i) off-pathway reactions can represent a serious obstacle on the protein folding road "down" to the native conformation, leading to increased levels of protein misfolding; ii) additional information supplementing the amino acid sequence and additional factors should probably be considered, as protein folding *in vivo* in many cases has been proven to be far more efficient than protein folding in a test tube [5,6]. Thus, comprehensive understanding of the mechanism of protein folding requires elucidation of the folding pathway under the native conditions, such as those that exist *in vivo* in the cell.

Folding *in vivo* takes place in a crowded cellular environment and thus is thought to be affected by a number of factors, such as viscosity of cytoplasm, local pH values, concentration and activity/affinity of surrounding macromolecules and the presence of folding catalysts and accessory proteins [5-7]. However, cellular components, such as chaperones and folding catalysts have been shown to be involved in kinetic partitioning between proper folding and aggregation; thus, they affect the yield of the correctly folded protein rather than the folding mechanism [5-9]. Therefore, elucidation of the mechanism of *in vivo* protein folding requires consideration of some additional factors and aspects of the *in vivo* process.

For instance, the following questions should be considered in this context:

1) How early during the process of protein synthesis *in vivo* can the folding as well as misfolding start/occur?
2) What additional factors *in vivo* in the cell can affect the protein folding mechanism?

This minireview will try to answer these and other questions dealing with the problem of *in vivo* protein folding and misfolding and focusing on the hypothesis suggesting that translation rates can affect the mechanism of *in vivo* protein folding leading to increased levels of misfolding.

2. PROTEIN FOLDING *IN VIVO* – CO-TRANSLATIONAL FOLDING

In vivo, protein synthesis is taking place on the ribosomes and in the early 60's and mid-late 70's the first observations came [10-16] suggesting that protein folding *in vivo* is largely a co-translational process. The idea was revisited in the late 80s and 90's, with many more substantial experiments to support the co-translational nature of the *in vivo* folding process [reviewed in 17-19]. Evidence suggesting the co-translational protein folding has mainly come from experiments that demonstrated (i) the enzymatic activity of the growing polypeptide on the ribosome [10-12; 20-23], (ii) the formation of correct epitopes able to bind corresponding conformational antibodies [13, 24-26], (iii) the formation of correct disulfide cross-bridges in the growing nascent chains [14-16, 25, 27-29], (iv) the formation of compact structures in the growing nascent chains (characteristic for the folded protein), that can resist proteolytic digestion [30-34], (v) the ability of co-factors and ligands (such as heme) to bind the growing polypeptide chain, (an indication that a conformation competent for binding has been achieved) [25, 34-38], (vi) the ability of the nascent chain to form oligomeric complexes with other polypeptides [11,12,15,16, 39-41] and (vii) quite recently, from experiments that assessed the distance between the amino acids in the growing polypeptide chain by the use of the Fluorescence Resonance Energy Transfer (FRET) technique [42; see 43 for a review].

These and other experiments have formulated the basic principles of *in vivo* (co-translational) protein folding, which are briefly summarized below:

1) *In vivo* protein folding is a vectorial process; i.e. the polypeptide chain is synthesized and folded from the N-terminal to the C-terminal end.
2) Attachment of the nascent polypeptide chain to the ribosome during the process of protein synthesis (through its C-terminal end) reduces the conformational space and the degrees of freedom of the growing chain.
3) *In vivo* protein folding starts very early during the process of polypeptide chain synthesis on the ribosome, with some elements (such as alpha-helices) forming inside the ribosomal tunnel.
4) Co-translational - "vectorial" nature of the folding of the nascent polypeptide from its N- to C-terminal end results in a sequential N- to C-terminal co-translational folding pathway and proceeds through the sequential structuring of the distinct regions of the polypeptide (i.e. subdomains and domains).
5) Certain chaperones and accessory proteins can bind the nascent chains and assist the co-translational folding as soon as the nascent chain emerges from the ribosome.
6) The vectorial nature of *in vivo* protein folding might be decisive for the folding mechanism as well as the pathway.

With acceptamce of the above mentioned principles an important question arises, namely, whether the rates of chain elongation can affect *in vivo* protein folding and thus influence intracellular protein misfolding. But why this might be important at all? Further, what is the significance of the potential link between the translation rates and the protein foldin?

3. Protein Elongation Rates Are Not Uniform

Protein synthesis in pro- and eukaryotes is a complex process which consists of three main steps: initiation (whereby the ribosome locates the initiation codon), elongation (which proceeds after initiation and results in actual synthesis of the polypeptide chain) and termination (leading to the release of the completed polypeptide chain from the ribosome).

It should be noted that rates of chain elongation during protein synthesis are not uniform [see 44,45 for ref.]. The initial observations demonstrating the discontinuous elongation rates came mostly from the seminal publications by Morris and colleagues, who showed that both eukaryotic (alpha- and beta-globin) and prokaryotic (MS2 coat protein) proteins are translated at rates that are not constant and that the translation of the above mentioned proteins proceeds more rapidly at some regions of the mRNA and more slowly at the others [46-48]. Subsequently, the phenomenon has been demonstrated for many other proteins/mRNA from multiple organisms [see 44,45 for ref.].

It was originally proposed that both the mRNA secondary structure [48] and codon usage [49,50] contribute to the non-uniformity of the translational elongation rate. However, subsequent careful analysis has demonstrated that after translational initiation (particularly in eukaryotes) the ribosome can, in most cases, locally destabilize secondary structures and move along the message without any significant delays [51]. This argues that codon usage is the primary cause of nonuniformity in translational elongation. Although several other additional factors have been also suggested to contribute to discontinuous elongation rates [52] (and for example, very complex secondary structures (e.g. those involving pseudoknots) can stall elongating ribosomes producing frameshifts [53] or significant reductions in protein expression levels [54]), it is now widely assumed that the rate of protein synthesis on the ribosome is predominantly modulated by non-random use of synonymous codons and the availability of the respective isoacceptor tRNAs [for ref. see 44,45,52]. Synonymous codons are utilized with different frequencies and a strong codon bias exists within the given organism [55-57]. The abundance of tRNAs is known to be directly proportional to the frequency of codon usage characteristic for a given organism [58], implying that a given frequent codon would be translated faster than an infrequent one. This postulate has been supported by numerous experiments involving either the change of synonymous frequent codons into infrequent codons (or vise versa) or the change in the level of expression/abundance of the certain tRNAs [49,50, 59-64 and ref. therein]. Thus, discontinuous elongation rates lead to an increased residence time of a ribosome at particular positions along mRNA. This means that different peptide bonds are formed at different rates. Would this in turn be important or would this influence the mechanism of co-translational protein folding?

4. Synonymous Codon Usage might Define the Pathway of Co-Translational Protein Folding

In the late 80's the group of Alistair Brown at Glasgow University, UK [64] and we [65] suggested that sequential folding events, which can take place during the co-translational folding of proteins might be separated by translational pauses and such regions of slowed down translation might serve as interpunctuations during co-translational protein folding. It was hypothesized that such optimization of the local translation rates might be necessary to fine tune the synthesis and folding of the nascent polypeptide chain growing on the ribosome, thus ensuring high accuracy of the *in vivo* co-translational folding process [64,65]. Thus, altered protein translation kinetics might lead to incorrect protein folding.

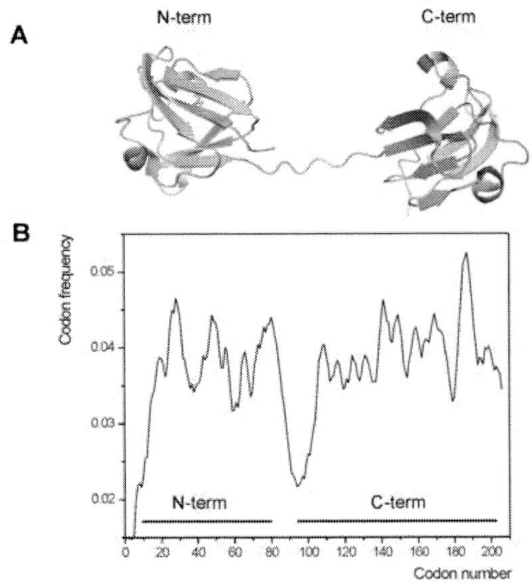

Figure 1. Backbone structure of the two domain protein beta-B2 crystallin and the codon frequency profile [44, 65-68]. A) Cartoon representation of the bovine (*Bos taurus*) beta-B2 crystallin structure (PDB ID 2BB2). B) Codon frequency profile (N- and C-terminal domain boundaries are indicated).

Is there, however, any evidence suggesting that rare synonymous codons/translational pauses might indeed serve as inter-punctuations marks/events during co-translational folding? A number of approaches have been undertaken to answer this question. First, attempts have been made to analyze the location of rare codons, especially clusters of rare codons with respect to the structure of the corresponding encoded protein regions [44,64-70]. The results indicated that protein domain boundaries that mark higher-order structural organization units are largely encoded by the rare codons and thus should represent slowly translated regions on the mRNA (Figures 1,2 and [64,66,68,69]). Moreover, in certain cases information has been obtained indicating that rare synonymous codons might also encode termini of the secondary and super-secondary structural elements [44,65,67,70]. Relative appearances of the most rare as well as frequent codons in investigated gene sequences were shown to display a very high degree of conservation [67]. It has also been found that structurally homologous proteins from different organisms display very similar codon frequency distribution profiles, despite significant variations in the codon usage spectra between the organisms (Figure 2 and [67]). This might mean that the codon frequency can be conserved at the particular position of mRNA, rather than a specific codon/amino acid, providing further support to the hypothesis. Genome wide analysis of mRNA/proteins with known sequences and 3D structures from pro- and eukaryotic organisms revealed intriguing correlations between the usage of synonymous codons and protein structural elements [71-80]. Synonymous codons showed non-randomness in their usage in various secondary structure elements [71-80]. Thus, there is apparently a selective pressure to keep a defined repertoire of synonymous codons for certain genes/proteins and keep them at selective locations/sites of their mRNAs. Could this be considered as a proof of the hypothesis, or it is possible that non random use of synonymous codons serves another purpose as well?

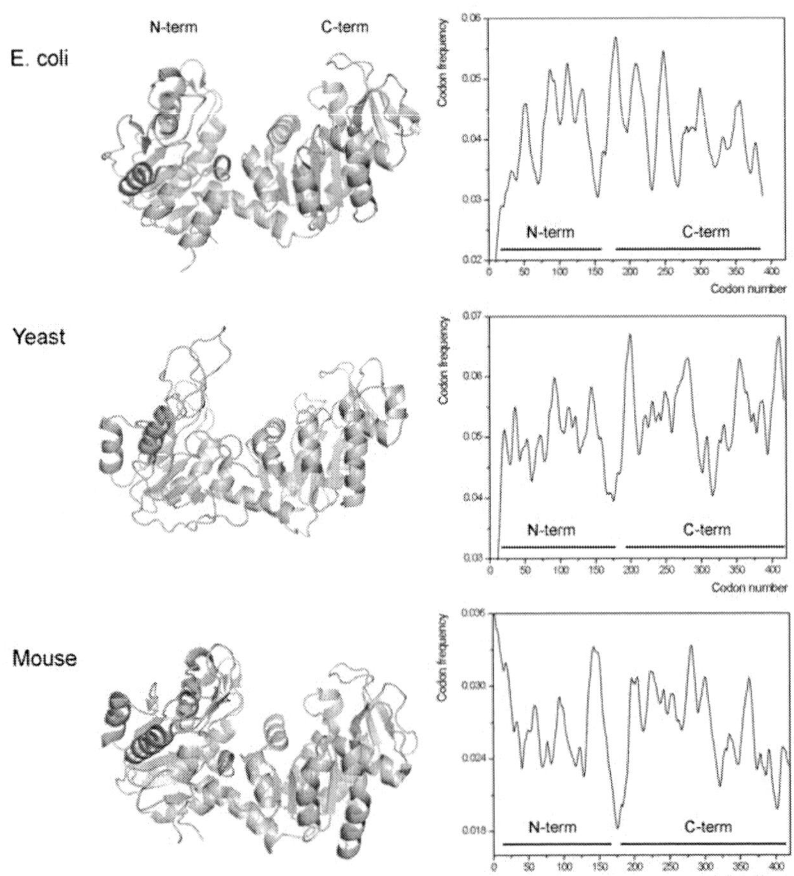

Figure 2. Backbone structure of the phoshpoglycerate kinases (PGK) from different organisms and the codon frequency profile. A) Cartoon representation of the *Escherichia coli* PGK (PDB ID 1ZMR) structure and the codon frequency profile on the right (N- and C-terminal domain boundaries are indicated). B) Cartoon representation of the yeast *Saccharomyces cerevisiae* PGK (PDB ID 3PGK) and the codon frequency profile on the right (N- and C-terminal domain boundaries are indicated). C) Cartoon representation of the mouse *Mus musculus* PGK (PDB ID 2P9T) and the codon frequency profile on the right (N- and C-terminal domain boundaries are indicated).

It should be noted in this connection that different proteins are required in cells in different quantities and one of the ways to modulate the level of protein expression is through the choice of synonymous codons [56, 81-86]. The enrichment of frequent/preferred synonymous codons in highly expressed genes is well documented [56, 81-86] and could provide the basis for one of the explanations. Yet, this would not explain why there is a pressure to keep certain synonymous codons at selective sites of mRNA (Figure 2 and [67,71-80]).

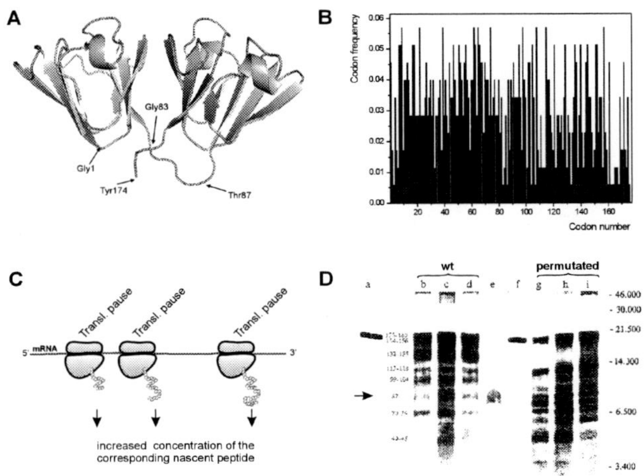

Figure 3. Backbone structure of the two domain protein gamma-B crystalline, codon frequency profile and PAGE of the products of cell-free translation (from [68]). A) Cartoon representation of the bovine (*Bos taurus*) gamma-B crystallin structure (PDB ID 4GCR). B) Codon frequency histogram. C) Schematic representation of the paused ribosomes. Increased residence time of a ribosome at particular positions along mRNA leads to increased accumulation of the nascent chains of the respective sizes. D) ^{35}S-Autoradiograph of PAGE of translation products (from [68]), a = gamma-B crystallin after 15 min translation; b-d = nascent peptides prepared from polyribosome fraction after 5, 10 and 15 min of gamma-B translation, respectively; e = gamma-B N-terminal domain translated separately; f = gamma circularly permuated protein (gamma-CP) after 15 min translation; g-i = nascent peptides prepared from polyribosome fraction after 5, 10 and 15 min of gamma-CP translation. Major peptides are denoted with numbers according to their lengths (estimated from peptide molecular weights). Arrow indicates the pause in the interdomain (linker region).

As protein folding is a modular process [1-3, 87-89], one might thus suggest that optimization of *in vivo* protein folding might include the adaptation of codon selection along mRNA to a particular translation kinetics [44,45, 64-70]. This could be in fact necessary to ensure step-by-step synthesis and folding of defined portions of the polypeptide chain growing on the ribosome from its N-terminal end and could explain the need for location of the rare synonymous codons at specific sites on the mRNA. It should be however noted that the folding of the nascent polypeptide chain under any circumstances will be governed by the physico-chemical properties of the amino acids forming the chain. Nevertheless, temporal separation of folding events by translation pauses might significantly help optimize the process and channel it to the most productive pathway [90]. Interestingly, one can also observe a certain hierarchy in the location of rare synonymous codons, with the rarest ones encoding domain boundaries and less rare ones encoding boundaries of subdomain regions and the respective boundaries of secondary structure elements [65-68]. This might reflect the need to provide a more substantial translational delay required for independent folding of larger structural units in comparison with smaller ones [65-68]. Although, the latter has not been substantially well documented, strong translation pauses have been indeed observed during the synthesis of domain and subdomain linkers (Figure 3 and [68]). Somewhat weaker translational delays have been observed during the synthesis of loops and turns connecting smaller structural units [44,68]. Particularly, as a consequence of the irregular distribution of rare codons along the polypeptide chain of gamma-B crystallin, translation of this two-domain protein was shown to be a non-uniform process characterized by specific pauses (Figure 3 and [68]). One of the major delays was shown to occur during translation of the connecting peptide between domains [68]. Analysis of the codon distribution along the gamma-B crystallin mRNA also revealed that the relative frequencies of codons in the N-terminal domain are about 2.5-fold higher than those of the C-terminal domain (Figure 3 and [68]). From this observation, one would also predict that the translation of gamma-B crystallin proceeds in a manner where the N-terminal domain would be translated faster than the C-terminal one. This was shown to be indeed the case [68]. The size distribution of nascent peptides has demonstrated that most of the pauses occur during the translation of the C-terminal domain of the natural gamma-B crystallin (Figure 3 and [68]). Independent evidence for this observation was obtained from translation experiments using the circularly permutated variant of gamma-B crystallin. In this construct, the N-terminal 81 residues of gamma-B crystallin are connected via a pentaglycine linker to the 3'-end of the C-terminal domain (residues 88-174) [68]. As has been shown (Figure 3 and [68]), in the latter case the pausing regions are shifted to the 5'-part of the mRNA, resulting in the predominant accumulation of nascent peptides with sizes smaller than half the size of the full-length protein. Obviously, this indicates that the C-terminal domain of gamma-B crystallin is translated at a slower rate compared to the N-terminal one, independent of its position within the complete gamma-B crystallin. Evidently, the relative residence time of ribosomes during translation of the C terminal-part of gamma-B mRNA is higher than that of the 5'-end of the mRNA [68]. Would this be significant for the co-translational folding of this molecule? Gamma-B crystallin from calf eye lens consists of two homologous domains connected by a six-residues linker peptide [Figure 3 and [91] for review]. Both sequence homology and structure topology of the two domains and the Greek key motifs suggest that the protein is the product of gene duplication [91]. However, in spite of the extreme symmetry of the molecule, its domains show significant differences in their stability against urea denaturation [92-94]. Unfolding/folding experiments reveal sequential and independent domain folding, with the C-terminal domain showing relatively low stability and enhanced unfolding compared to the N-terminal domain which requires extreme denaturant conditions to unfold [92-94]. These findings suggest that the N-terminal domain serves as a core structure for the folding and stabilization of the C-terminal half of the molecule, thus generating the anomalously high stability of the gamma-B crystallin [91-94]. The translation experiments give clear evidence that the modular structure of gamma-B-crystallin is reflected by non-uniform translation kinetics, with pauses in the subdomain and domain regions of the mRNA [68]. The overall rate of translation of the N-terminal domain exceeds the translation of the C-terminal domain, to the effect that the N-terminal half of the molecule may serve to stabilize the marginally stable C-terminal half [68]. This suggests that the translation rates in the case of gamma-B crystallin can be indeed optimized to tune the synthesis and co-translational folding of the nascent polypeptide chain [68]. Thus, non-uniform translation and sequential domain folding might represent two different mechanisms to optimize co-translational modular selfassembly of domain proteins. It has been also suggested that ribosome pausing may in addition facilitate co-translational binding of co-factors [35,37,38,44] and aid the integration of transmembrane segments of membrane proteins into membranes [37,38].

5. Co-Translational Protein Misfolding, the Effect of Synonymous Codon Usage and Translation Rates

Cells have developed numerous, sophisticated quality-control mechanisms to get rid of abnormally folded and misfolded proteins [see 95 for review]. Proteins that escape the cellular quality-control mechanisms can either be degraded or alternatively might form insoluble aggregates resisting degradation and subsequently causing a number of so-called "conformational" or "misfolding" diseases in humans and animals [96]. Abnormal function or dis-regulation of either the degradation machinery or the quality control systems leads to increased levels of misfolding due to a shift in the balance between the levels of the protein produced in the cell and the components of the above mentioned control systems and degradation machineries [95,96]. Yet, these systems *per se* could not be the cause of misfolding. Oxidative and temperatures stresses, changes in the pH as well as other factors are known to be the primary cause of cellular misfolding [97-99]. Could there be additional factors that might affect the intracellular folding pathway and the mechanism? The phenomenon of co-translational ubiquitination (and degradation) of proteins [100-102] that is observed under non-stress conditions suggests that i) misfolding can start very early during protein synthesis on the ribosome and that ii) there might be additional factors that could influence the mechanism of protein folding in the cell during protein translation. So, what can go wrong during protein synthesis and co-translational folding of proteins?

As we hypothesized above altered translation rates may lead to increased levels of protein misfolding. Temporal separation of folding events during protein synthesis on the ribosome might be necessary to avoid "unwanted" interactions within the growing polypeptide. It can be suggested that abnormal (e.g. accelerated) rates of translation at a certain region of the mRNA would allow particular parts of the polypeptide chain to appear earlier in time during translation and this could affect interaction of this polypeptide region with the preceding one, which (one may speculate) was not yet properly folded (due to a lack of time) (Figure 4.). Such misfolded nascent chains can further undergo enhanced interaction (co- or posttranslational) with chaperones (due to their altered chaperone affinity) and this could shift the equilibrium back to the productive folding pathway. Alternatively, such misfolded nascent chains could be trapped and degraded or form insoluble aggregates that my lead to amyloid formation and many diseases (Figure 4).

So, what factors can affect the local rates of translation? Translation rates of individual codons might differ substantially (5 to 25-fold) [103,104] and as has been mentioned above mainly depend on the concentration/levels of cognate aminoacyl-tRNAs surrounding the ribosome during the translation. It is assumed that acylated tRNAs, which surround the ribosome, randomly encounter the A site (in complex with elongation factor 1A (eEF1A) and GTP) and become temporally trapped. Trapped species is then tested for proper pairing with the corresponding codon in the A site [105]. If the pairing is correct, transpeptidation reaction becomes irreversible, if not, the ternary (eEF1A•GTP•aminoacyl-tRNA) complex is then rejected and another complex is processed in turn [105]. Several factors might affect the translation rates of individual codons and the local translation rates of the mRNAs. These include factors that affect the levels of expression of various tRNAs in the cell, the efficiency of their aminoacylation as well as rates at which they are selected by the ribosome. In addition, silent synonymous codon substitutions/polymorphisms changing e.g. infrequent to frequent codons and vise versa can also affect the local translations rates. But would this indeed lead to protein misfolding or any change in the protein structure?

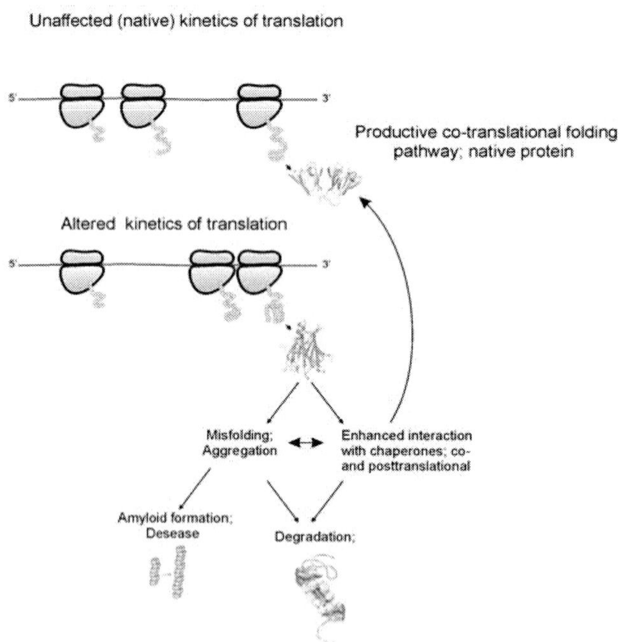

Figure 4. Schematic representation of the possible influence of translation kinetics on co-translational protein folding and misfolding (modified from [45]).

Soon after the hypothesis had been formulated, attempts (although very limited) have been made to prove it experimentally, employing artificial silent mutagenesis of e.g. infrequent to frequent codons [45,106,107]. Both, *in vitro* cell-free protein translation systems as well as *in vivo* cell systems have been used. These experiments provided evidence in support of the hypothesis [45,106,107]. Synonymous codon replacements in *E. coli* chloramphenicol acetyltransferase gene have led to a decreased specific activity of the enzyme synthesized in an *in vitro* cell-free system [45]. Similar experiments involving *in vivo* cell systems and mutagenesis of synonymous codons in the yeast enzyme anthranilate synthase:indole-3-glycerol phosphate synthase expressed in *Saccharomyces cerevisiae* [106] and *Echinococcus granulosus* fatty acid binding protein expressed in *E. coli* [107] resulted in altered specific activities of both proteins. Since the specific activity of a given protein could be considered as a measure of its proper folding, one might conclude that proteins produced from the silently mutated mRNAs were partially misfolded. But, would there be any natural example in support of the idea?

Very recently, Kimchi-Sarfaty and colleagues provided evidence that naturally occurring silent SNPs can affect *in vivo* protein folding and, consequently, function [108]. Kimchi-Sarfaty and colleagues showed that substrate specificity of P-glycoprotein (Pg-p), the product of the multidrug resistance 1 (*MDR1*) gene, is altered by single nucleotide polymorphisms presumed to be synonymous and silent. The authors suggested that the C3435T polymorphism that changes the frequent ATC codon to a less frequent synonymous ATT codon at position 1145 (both encode isoleucine) of the *MDR1* open reading frame is primarily responsible for the change in activity of the P-glycoprotein [108]. They were further able to reveal conformational differences between the wild-type and the P-gp variants carrying silent synonymous codon substitutions and suggest that these differences could account for the observed difference in protein's substrate specificity [108]. Although it is not quite clear what step in the P-gp folding process has been affected by the silent SNPs under investigation [90,108,109], one might suggest that slowing down of ribosome traffic at a selected mRNA region might have not allowed certain intermediate(s) to correctly fold (and/or e.g. be properly inserted into the membrane; P-gp is a transmembrane protein) and this might have affected the P-gp folding pathway and thus the final conformation of a protein. Interestingly, ATC and ATT isoleucine codons show different (and statistically significant) preferences for different secondary structure types in human proteins [71]. Further experiments are required to elucidate which particular step on the P-gp folding pathway might have been affected. However, it should be noted that it has been previously shown that ribosome pausing may indeed play a critical role in the co-translational folding of another transmembrane protein -- photosynthetic reaction center protein D1 [38,110]. Translation pausing in the case of D1 was suggested to define the timing of co-factors (such as chlorophyll) binding to D1 and aid in the integration of D1 into thylakoid membranes [38,110]. In any circumstance, the study of Kimchi-Sarfaty and colleagues [108] is the first publication demonstrating that naturally occurring silent SNPs might affect protein folding and activity without changing mRNA and protein levels. Would there be other evidence in support of the hypothesis?

In recent years, the demand for recombinant proteins used in research laboratories or in medical settings (e.g. for therapeutic applications) has increased dramatically. Thus, there is an evident and urgent need for the development of efficient, reliable and cost effective methods/ways to produce large amounts of recombinant proteins. As has been mentioned above, a strong codon bias exists within many organisms [55-57]. Individual synonymous codons appeared to be utilized with very different and sometimes opposite frequencies between different organisms [55-57]. It is well known that, for example, many frequent synonymous codons in yeast *Sacchoromyces cerevisiae* as well as mammals (e.g. Arg-AGA/AGG; Ile-ATA) are utilized with low frequency in *Escherichia coli* [56,57] – a most common host for heterologous protein production. Since the effect of synonymous codon substitutions on overall levels of protein expression is widely recognized and well documented [56, 60-63, 81-86], it is therefore possible to modulate the level of protein expression by suitable selection of synonymous codons [60-63]. At present, the majority of biotech companies and research laboratories tend to change (in the target gene) most of the infrequent synonymous codons to the frequent ones with the hope to boost the protein production levels. Indeed, this approach often yields large amounts of recombinant proteins produced [111-114 and ref therein]. However, it turns out that optimization of heterologous gene expression by a choice of only frequent codons often yields biologically inactive products forming insoluble aggregates, which have to be renatured artificially in order to regain similarity in structure and biological activity with native analogues [112,114-118]. These effects could be due to the affected mechanism of co-translational protein folding. Moreover, in many instances such recombinant proteins also appeared to be abnormally processed. It was for example found that introduction of non-optimal codons in the signal sequences of secretory proteins exported via the sec-dependent pathway in *E. coli* may lead to both protein misfolding and premature protein degradation [119,120]. It has been hypothesized that codon context within the signal sequence of the *E. coli* secretory proteins has evolved to play a key role in protein export by introducing a translational pause to facilitate folding of the emerging nascent chain and its proper interaction with the components of the export machinery [119,120]. The importance of pausing was also demonstrated in the case of the signal recognition particle receptor alpha-subunit co-translational assembly and membrane binding (in eukaryotic rabbit reticulocyte lysate cell-free translation system) [121].

Undoubtedly, the knowledge that synonymous codon substitutions might affect protein folding, stability, and activity should not be neglected in attempts to design synthetic genes or optimize the expression of natural genes for heterologous protein production. This knowledge should be also taken into account in assessing the likelihood of the development and progression of many misfolding diseases (especially those that might be linked to silent SNPs) as well as in the complete elucidation of the mechanism of *in vivo* co-translational protein folding. In this context, it ought to be mentioned that first attempts to build *in silico* models of co-translational protein folding that would take into account the kinetics of protein translation gave very promising results [122]. Using simple hydrophobic-polar (HP) lattice models, Huard and colleagues have clearly demonstrated that the kinetics of protein translation has a significant impact on the final conformation of the protein during the co-translational folding [122]. It has been shown that sequential ground states become less numerous and more compact in cases where co-translational folding is proceeding under kinetic control [122].

Although much of the attention above has been devoted to discussing of the role of rare synonymous codons, it ought to be underlined that frequent codons should be considered equally important in the process of co-translational protein folding. Unique characteristics of the distribution of both rare and frequent codons along the mRNA define local rates of translation and specify overall features of the kinetics of translation for different mRNAs. Therefore, distribution of both frequent and rare codons in mRNA should be considered as a unique kinetic guide for co-translational folding, driving the spatial packing of the synthesized polypeptide chain. One might suggest that, both in eukaryotic and prokaryotic cells, specific features of non-uniform translation could be necessary to ensure proper in vivo protein folding and a low level of protein misfolding.

ACKNOWLEDGMENTS

The author would like to thank Drs, Crystal Weyman and Barsanjit Mazumder for the critical reading of the manuscript and helpful discussions.

REFERENCES

[1] Jaenicke, R., & Lilie, H. (2000). Folding and association of oligomeric and multimeric proteins. *Adv. Protein Chem., 53,* 329-401.

[2] Onuchic, J.N., & Wolynes, P.G. (2004). Theory of protein folding. *Curr. Opin. Struct. Biol., 14,* 70-75.

[3] Lindberg, M.O., & Oliveberg, M. (2007). Malleability of protein folding pathways: a simple reason for complex behaviour. *Curr. Opin. Struct. Biol., 17,* 21-29.

[4] Anfinsen, C.B. (1973). Principles that govern the folding of protein chains. *Science, 181,* 223-230.

[5] Ellis, R.J. (1996). Revisiting the Anfinsen cage. *Fold. Des., 1,* R9-15.

[6] Naylor, D.J., & Hartl, F.U. (2001) Contribution of molecular chaperones to protein folding in the cytoplasm of prokaryotic and eukaryotic cells. *Biochem. Soc. Symp., 68,* 45-68.

[7] Frydman, J. (2001). Folding of newly translated proteins in vivo: the role of molecular chaperones. *Annu. Rev. Biochem., 70,* 603-647.

[8] Jaenicke, R. (1998). Protein self-organization in vitro and in vivo: partitioning between physical biochemistry and cell biology. *Biol. Chem., 379,* 237-243.

[9] Thirumalai, D., & Lorimer, G.H. (2001). Chaperonin-mediated protein folding. *Annu. Rev. Biophys. Biomol. Struct., 30,* 245-269.

[10] Cowie, D. B., Spiegelman, S., Roberts, R. B., & Duerksen, J. D. (1961). Ribosome-bound β-galactosidase. *Proc. Natl. Acad. Sci. U. S. A., 47,* 114-122.

[11] Zipser, D., & Perrin, D. (1963). Complementation on ribosomes. *Cold Spring Harbor Symp. Quant. Biol., 28,* 533-537.

[12] Kiho, Y., & Rich, A. (1964). Induced Enzyme Formed on Bacterial Polyribosomes. *Proc. Natl. Acad. Sci. U. S. A., 51,* 111-118.

[13] Hamlin, J., & Zabin, I. (1972). β-Galactosidase: Immunological Activity of Ribosome-Bound, Growing Polypeptide Chains. *Proc. Natl. Acad. Sci. U. S. A., 69,* 412-416.

[14] Bergman, L.W., & Kuehl, W.M. (1979). Formation of intermolecular disulfide bonds on nascent immunoglobulin polypeptides. *J. Biol. Chem., 254,* 5690-5694.

[15] Bergman, L.W., & Kuehl, W.M. (1979). Formation of an intrachain disulfide bond on nascent immunoglobulin light chains *J. Biol. Chem., 254,* 8869-8876.

[16] Bergman, L.W., & Kuehl, W.M. (1979). Co-translational modification of nascent immunoglobulin heavy and light chains. *J. Supramol. Struct. 11,* 9-24.

[17] Fedorov, A.N., & Baldwin, T.O. (1997). Cotranslational protein folding. *J. Biol. Chem. 272,* 32715-32718.

[18] Hardesty, B., & Kramer, G. (2001). Folding of a nascent peptide on the ribosome. *Prog. Nucleic Acid Res. Mol. Biol. 66,* 41-66.

[19] Kolb, V.A. (2001). Cotranslational protein folding. *Mol. Biol. (Mosk.), 35*, 682-690.

[20] Kolb, V.A., Makeyev, E.V., & Spirin, A.S. (1994). Folding of firefly luciferase during translation in a cell-free system. *EMBO J., 13,* 3631-3637.

[21] Kudlicki, W., Chirgwin, J., Kramer, G., & Hardesty, B. (1995). Folding of an enzyme into an active conformation while bound as peptidyl-tRNA to the ribosome. *Biochemistry, 34*, 14284-14287.

[22] Makeyev, E.V., Kolb, V.A., & Spirin, A.S. (1996). Enzymatic activity of the ribosome-bound nascent polypeptide. *FEBS Lett. 378*, 166-170.

[23] Kolb, V.A., Makeyev, E.V., & Spirin, A.S. (2000). Co-translational folding of an eukaryotic multidomain protein in a prokaryotic translation system. *J. Biol. Chem. 275*, 16597-16601.

[24] Friguet, B., Djavadi-Ohaniance, L., King, J., & Goldberg, M.E. (1994). In vitro and ribosome-bound folding intermediates of P22 tailspike protein detected with monoclonal antibodies. *J. Biol. Chem., 269*, 15945-15949.

[25] Chen, W., Helenius, J., Braakman, I., & Helenius, A. (1995). Cotranslational folding and calnexin binding during glycoprotein synthesis. *Proc. Natl. Acad. Sci. U. S. A., 92*, 6229-6233.

[26] Clark, P.L., & King, J. (2001). A newly synthesized, ribosome-bound polypeptide chain adopts conformations dissimilar from early in vitro refolding intermediates. *J. Biol. Chem., 276*, 25411-25420.

[27] Peters, T., & Davidson, L. K. (1982). The biosynthesis of rat serum albumin. In vivo studies on the formation of the disulfide bonds. *J. Biol. Chem., 257,* 8847-8853.

[28] Pollitt, S., & Zalkin, H. (1983). Role of primary structure and disulfide bond formation in beta-lactamase secretion. *J. Bacteriol., 153*, 27-32.

[29] Kleizen, B., van Vlijmen. T., de Jonge, H.R., & Braakman, I. (2005). Folding of CFTR is predominantly cotranslational. *Mol. Cell., 20*, 277-287.

[30] Frydman, J., Nimmesgern, E., Ohtsuka, K., & Hartl, F.U. (1994). Folding of nascent polypeptide chains in a high molecular mass assembly with molecular chaperones. *Nature, 370*, 111-117.

[31] Netzer, W.J., & Hartl, F.U. (1997). Recombination of protein domains facilitated by co-translational folding in eukaryotes. *Nature, 388*, 343-349.

[32] Lin, L., DeMartino, G.N., & Greene, W.C. (1998). Cotranslational biogenesis of NF-kappaB p50 by the 26S proteasome. *Cell, 92*, 819-828.

[33] Nicola, A.V., Chen, W., & Helenius, A. (1999). Co-translational folding of an alphavirus capsid protein in the cytosol of living cells. *Nat. Cell. Biol., 6*, 341-345.

[34] Frydman, J., Erdjument-Bromage, H., Tempst, P., & Hartl, F.U. (1999). Co-translational domain folding as the structural basis for the rapid de novo folding of firefly luciferase. *Nat. Struct. Biol., 6,* 697-705.

[35] Komar, A.A., Kommer, A., Krasheninnikov, I.A., & Spirin, A.S. (1993). Cotranslational heme binding to nascent globin chains. *FEBS Lett., 326,* 261-263.

[36] Komar, A.A., Kommer, A., Krasheninnikov, I.A., & Spirin, A.S. (1997). Cotranslational folding of globin. *J. Biol. Chem., 272,* 10646-10651.

[37] Mullet, J.E., Klein, P.G., & Klein, R.R. (1990). Chlorophyll regulates accumulation of the plastid-encoded chlorophyll apoproteins CP43 and D1 by increasing apoprotein stability. *Proc. Natl. Acad. Sci. U. S. A., 87,* 4038-4042.

[38] Kim, J., Klein, P.G., & Mullet, J.E. (1991). Ribosomes pause at specific sites during synthesis of membrane-bound chloroplast reaction center protein D1. *J. Biol. Chem., 266,* 14931-14938.

[39] Gilmore, R., Coffey, M.C., Leone, G., McLure, K., & Lee, P.W.K. (1996). Co-translational trimerization of the reovirus cell attachment protein. *EMBO J., 15,* 2651-2658.

[40] Fedorov, A.N., & Baldwin T.O. (1999). Process of biosynthetic protein folding determines the rapid formation of native structure. *J. Mol. Biol., 294*, 579-586.

[41] Lin, L, DeMartino, G.N., & Greene W.C. (2000). Cotranslational dimerization of the Rel homology domain of NF-kappaB1 generates p50-p105 heterodimers and is required for effective p50 production. *EMBO J., 19*, 4712-4722.

[42] Woolhead, C.A, McCormick, P.J., & Johnson, A.E. (2004). Nascent membrane and secretory proteins differ in FRET-detected folding far inside the ribosome and in their exposure to ribosomal proteins. *Cell, 116,* 725-736.

[43] Johnson, A.E. (2005). The co-translational folding and interactions of nascent protein chains: a new approach using fluorescence resonance energy transfer. *FEBS Lett. 579,* 916-920.

[44] Krasheninnikov, I.A., Komar, A.A. & Adzhubei I.A. (1991). Nonuniform size distribution of nascent globin peptides, evidence for pause localization sites, and a contranslational protein-folding model. *J. Protein Chem., 10*, 445–454.

[45] Komar, A.A, Lesnik, T. & Reiss C. (1999). Synonymous codon substitutions affect ribosome traffic and protein folding during in vitro translation. *FEBS Lett., 462*, 387-391.

[46] Protzel, A., & Morris, A.J. (1974). Gel chromatographic analysis of nascent globin chains. Evidence of nonuniform size distribution. *J. Biol. Chem., 249,* 4594-4600.

[47] Chaney, W.G., & Morris, A.J. (1978). Nonuniform size distribution of nascent peptides: the role of messenger RNA. *Arch. Biochem. Biophys., 191*, 734-741.

[48] Chaney, W.G., & Morris, A.J. (1979). Nonuniform size distribution of nascent peptides. The effect of messenger RNA structure upon the rate of translation. *Arch. Biochem. Biophys., 194,* 283-291.

[49] Pedersen, S. (1984). *Escherichia coli* ribosomes translate in vivo with variable rate. *EMBO J., 3*, 2895-2898.

[50] Bonekamp, F., Andersen, H.D., Christensen, T., & Jensen, K.F. (1985). Codon-defined ribosomal pausing in *Escherichia coli* detected by using the pyrE attenuator to probe the coupling between transcription and translation. *Nucleic Acids Res. 13*, 4113-4123.

[51] Liebhaber, S.A., Cash, F.E., & Shakin, S.H. (1984). Translationally associated helix-destabilizing activity in rabbit reticulocyte lysate. *J. Biol. Chem., 259*, 15597-15602.

[52] Buchan, J.R., & Stansfield, I. (2007). Halting a cellular production line: responses to ribosomal pausing during translation. *Biol. Cell., 99,* 475-487.

[53] Kontos, H., Napthine, S., & Brierley, I. (2001). Ribosomal pausing at a frameshifter RNA pseudoknot is sensitive to reading phase but shows little correlation with frameshift efficiency. *Mol Cell Biol., 21*, 8657-8670.

[54] Nackley, A.G., Shabalina, S.A., Tchivileva, I.E., Satterfield, K., Korchynskyi, O., Makarov, S.S., Maixner, W., & Diatchenko, L. (2006). Human catechol-O-methyltransferase haplotypes modulate protein expression by altering mRNA secondary structure. *Science, 314,* 1930-1933.

[55] Grantham, R., Gautier, C., Gouy, M., Mercier, R., & Pavé, A. (1980). Codon catalog usage and the genome hypothesis. *Nucleic Acids Res., 8*, r49-r62.

[56] Sharp, P.M., Tuohy, T.M., & Mosurski, K.R. (1986). Codon usage in yeast: cluster analysis clearly differentiates highly and lowly expressed genes. *Nucleic Acids Res. 14*, 5125-5143.

[57] Sharp, P.M., Cowe, E., Higgins, D.G., Shields, D.C., Wolfe, K.H., & Wright, F. (1988). Codon usage patterns in *Escherichia coli, Bacillus subtilis, Saccharomyces cerevisiae, Schizosaccharomyces pombe, Drosophila melanogaster and Homo sapiens*; a review of the considerable within-species diversity. *Nucleic. Acids Res. 16*, 8207-8211.

[58] Ikemura, T. (1985). Codon usage and tRNA content in unicellular and multicellular organisms. *Mol Biol Evol., 2*, 13-34.

[59] Candelas, G.C., Ortiz, A., & Ortiz, N.. (1989). Features of the cell-free translation of a spider fibroin mRNA. *Biochem Cell Biol., 67*, 173-176.

[60] Chen, G.T., & Inouye, M. (1994). Role of the AGA/AGG codons, the rarest codons in global gene expression in *Escherichia coli*. *Genes Dev., 8*, 2641-2652.

[61] Kane, J.F. (1995). Effects of rare codon clusters on high-level expression of heterologous proteins in *Escherichia coli*. *Curr. Opin. Biotechnol. 6,* 494-500.

[62] Del Tito, B.J. Jr, Ward, J.M., Hodgson, J., Gershater, C.J., Edwards, H., Wysocki, L.A., Watson, F.A., Sathe, G., & Kane, J.F. (1995). Effects of a minor isoleucyl tRNA on heterologous protein translation in *Escherichia coli*. *J. Bacteriol., 177,* 7086-7091.

[63] Komar, A.A., Guillemet, E., Reiss, C., & Cullin, C. (1998). Enhanced expression of the yeast Ure2 protein in *Escherichia coli*: the effect of synonymous codon substitutions at a selected place in the gene. *Biol Chem., 379*, 1295-1300.

[64] Purvis, I.J., Bettany, A.J., Santiago, T.C., Coggins, J.R., Duncan, K., Eason, R., & Brown, A.J. (1987). The efficiency of folding of some proteins is increased by controlled rates of translation in vivo. A hypothesis. *J Mol Biol. 193*, 413-417.

[65] Krasheninnikov, I.A., Komar, A.A., & Adzhubei, I.A. (1988) Role of the rare codon clusters in defining the boundaries of polypeptide chain regions with identical secondary structures in the process of co-translational folding of proteins. *Dokl Akad Nauk SSSR,* 303, 995-999.

[66] Krasheninnikov, I.A., Komar, A.A., & Adzhubei, I.A. (1989). Frequency of using codons in mRNA and coding of the domain structure of proteins. *Dokl Akad Nauk SSSR, 305,* 1006-1012.

[67] Krasheninnikov, I.A., Komar, A.A., & Adzhubei, I.A. (1989). Role of the code redundancy in determining cotranslational protein folding. *Biokhimiia*, *54,* 187-200.

[68] Komar, A.A., & Jaenicke, R. (1995). Kinetics of translation of gamma B crystallin and its circularly permutated variant in an in vitro cell-free system: possible relations to codon distribution and protein folding. *FEBS Lett., 376*, 195-198.

[69] Thanaraj, T.A., & Argos, P. (1996). Ribosome-mediated translational pause and protein domain organization. *Protein Sci., 5,* 1594-1612.

[70] Thanaraj, T.A., & Argos, P. (1996). Protein secondary structural types are differentially coded on messenger RNA. *Protein Sci. 5*, 1973-1983.

[71] Adzhubei, A.A., Adzhubei, I.A., Krasheninnikov, I.A. & Neidle, (1996). Non-random usage of 'degenerate' codons is related to protein three-dimensional structure. *FEBS Lett.*, 399, 78-82.

[72] Xie, T. & Ding, D. (1998). The relationship between synonymous codon usage and protein structure. *FEBS Lett. 434*, 93-96.

[73] Oresic, M, & Shalloway, D. (1998) Specific correlations between relative synonymous codon usage and protein secondary structure. *J Mol Biol. 281,* 31-48.

[74] Gupta, S.K., Majumdar, S., Bhattacharya, T.K., & Ghosh, T.C. (2000) Studies on the relationships between the synonymous codon usage and protein secondary structural units. *Biochem Biophys Res Commun. 269*, 692-696.

[75] Chiusano, M.L., Alvarez-Valin, F., Di Giulio, M., D'Onofrio, G,, Ammirato, G., Colonna, G., & Bernardi, G. (2000). Second codon positions of genes and the secondary structures of proteins. Relationships and implications for the origin of the genetic code. *Gene, 261*, 263-269.

[76] Oresic, M., Dehn, M., Korenblum, D., & Shalloway, D. (2003) Tracing specific synonymous codon-secondary structure correlations through evolution. *J. Mol. Evol.* 56, 473-484.

[77] Gu, W., Zhou, T., Ma, J., Sun, X., & Lu, Z. (2003). Folding type specific secondary structure propensities of synonymous codons. *IEEE Trans Nanobioscience, 2*, 150-157.

[78] Gu, W., Zhou, T., Ma, J., Sun, X., & Lu, Z. (2004). The relationship between synonymous codon usage and protein structure in *Escherichia coli* and *Homo sapiens. Biosystems, 73,* 89-97.

[79] Song, J, Wang, M, & Burrage, K. (2006). Exploring synonymous codon usage preferences of disulfide-bonded and non-disulfide bonded cysteines in the *E. coli* genome. *J. Theor. Biol., 241*, 390-401.

[80] Mukhopadhyay, P., Basak, S., & Ghosh, T.C. (2007). Synonymous codon usage in different protein secondary structural classes of human genes: implication for increased non-randomness of GC3 rich genes towards protein stability. *J Biosci., 32,* 947-963.

[81] Bennetzen, J.L., & Hall, B.D. (1982). Codon selection in yeast. *J. Biol. Chem. 257*, 3026-3031.

[82] Sharp, P.M, & Devine, K.M. (1989). Codon usage and gene expression level in *Dictyostelium discoideum*: highly expressed genes do 'prefer' optimal codons. *Nucleic. Acids Res. 17*, 5029-5039.

[83] Kurland, C.G. (1991). Codon bias and gene expression. *FEBS Lett., 285,* 165-169.

[84] Karlin, S., Mrázek, J., & Campbell, A.M. (1998) Codon usages in different gene classes of the *Escherichia coli* genome. *Mol Microbiol. 29,* 1341-1355.

[85] Jansen, R., Bussemaker, H.J., & Gerstein, M. (2003). Revisiting the codon adaptation index from a whole-genome perspective: analyzing the relationship between gene expression and codon occurrence in yeast using a variety of models. *Nucleic Acids Res., 31*, 2242-2251.

[86] Lavner, Y., & Kotlar, D. (2005). Codon bias as a factor in regulating expression via translation rate in the human genome. *Gene, 345*, 127-138.

[87] Wetlaufer, D.B. (1981). Folding of protein fragments. *Adv. Protein Chem. 34*, 61-92.

[88] Hockenmaier, J., Joshi, A.K., & Dill, K.A. (2007) Routes are trees: the parsing perspective on protein folding. *Proteins, 66,* 1-15.

[89] Brockwell, D.J., & Radford, S.E. (2007). Intermediates: ubiquitous species on folding energy landscapes? *Curr. Opin. Struct. Biol. 17,* 30-37.

[90] Komar, A.A. (2007). SNPs, silent but not invisible. *Science, 315,* 466-467.

[91] Bloemendal, H., de Jong, W., Jaenicke, R., Lubsen, N.H., Slingsby, C., & Tardieu, A. (2004). Ageing and vision: structure, stability and function of lens crystallins. *Prog. Biophys. Mol. Biol., 86*, 407-485.

[92] Rudolph, R., Siebendritt, R., Nesslauer, G., Sharma, A.K., & Jaenicke, R. (1990). Folding of an all-beta protein: independent domain folding in gamma II-crystallin from calf eye lens. *Proc. Natl. Acad. Sci. U. S. A., 87*, 4625-4629.

[93] Sharma, A.K., Minke-Gogl, V., Gohl, P., Siebendritt, R., Jaenicke, R., & Rudolph, R. (1990). Limited proteolysis of gamma II-crystallin from calf eye lens. Physicochemical studies on the N-terminal domain and the intact two-domain protein. *Eur. J. Biochem., 194*, 603-609.

[94] Mayr, E.M., Jaenicke, R., & Glockshuber, R. (1997). The domains in gamma-B crystallin: identical fold-different stabilities. *J. Mol. Biol., 269*, 260-269.

[95] Bukau, B., Weissman, J. & Horwich, A. (2006). Molecular chaperones and protein quality control. *Cell, 125*, 443-451.

[96] Chiti, F. & Dobson, C.M. (2006). Protein misfolding, functional amyloid, and human disease. *Annu. Rev. Biochem., 75*, 333–366.

[97] Macario, A.J., & Conway de Macario, E. (2000). Stress and molecular chaperones in disease. *Int J Clin Lab Res., 30*, 49-66.

[98] Bieschke, J., Zhang, Q., Bosco, D.A., Lerner, R.A., Powers, E.T., Wentworth, P. Jr, & Kelly, J.W. (2006). Small molecule oxidation products trigger disease-associated protein misfolding. *Acc. Chem. Res., 39*, 611-619.

[99] Malhotra, J.D., & Kaufman, R.J. (2007). Endoplasmic reticulum stress and oxidative stress: a vicious cycle or a double-edged sword? *Antioxid. Redox Signal., 9*, 2277-2293.

[100] Sato, S., Ward, C.L., & Kopito, R.R. (1998). Cotranslational ubiquitination of cystic fibrosis transmembrane conductance regulator in vitro. *J. Biol. Chem., 273*, 7189-7192.

[101] Zhou, M., Fisher, E.A., & Ginsberg, H.N. (1998). Regulated co-translational ubiquitination of apolipoprotein B100. A new paradigm for proteasomal degradation of a secretory protein. *J. Biol. Chem., 273*, 24649-24653.

[102] Lin, L., DeMartino, G.N., & Greene, W.C. (1998). Cotranslational biogenesis of NF-kappaB p50 by the 26S proteasome. *Cell, 92*, 819-828.

[103] Curran, J.F., & Yarus, M. (1989). Rates of aminoacyl-tRNA selection at 29 sense codons in vivo. *J. Mol. Biol., 209*, 65-77.

[104] Sorensen, M.A., & Pedersen, S. (1991). Absolute in vivo translation rates of individual codons in *Escherichia coli*. The two glutamic acid codons GAA and GAG are translated with a threefold difference in rate. *J. Mol. Biol. 222*, 265-280.

[105] Merrick, W.C. & Komar, A.A. (2004). Eukaryotic protein biosynthesis: The elongation cycle. In *Encyclopedia of Biological Chemistry* (Lennarz W.J. and Lane M. D., eds, Vol. 4, pp. 224-230), Oxford, Academic Press/Elsevier Science, Elsevier.

[106] Crombie, T., Boyle, J.P., Coggins, J.R. & Brown A.J. (1994). The folding of the bifunctional TRP3 protein in yeast is influenced by a translational pause which lies in a region of structural divergence with *Escherichia coli* indoleglycerol-phosphate synthase. *Eur. J. Biochem., 226*, 657-664.

[107] Cortazzo P., Cervenansky C., Marin M., Reiss C., Ehrlich R., and Deana A. (2002). Silent mutations affect in vivo protein folding in Escherichia coli. Biochem Biophys Res Commun. 293, 537-541.

[108] Kimchi-Sarfaty, C., Oh, J.M., Kim, I.W., Sauna, Z.E., Calcagno, A.M., Ambudkar S.V., & Gottesman M.M. (2007). A "silent" polymorphism in the MDR1 gene changes substrate specificity. *Science, 315*, 525-528.

[109] Komar, A.A. (2007). Silent SNPs; impact on gene function and phenotype. *Pharmacogenomics, 8*, 1075-1080.

[110] Kim, J., Klein, P.G., & Mullet, J.E. (1994). Synthesis and turnover of photosystem II reaction center protein D1. Ribosome pausing increases during chloroplast development. *J. Biol. Chem., 269*, 17918-17923.

[111] Gustafsson, C., Govindarajan, S., & Minshull, J. (2004). Codon bias and heterologous protein expression. *Trends Biotechnol., 22*, 346-353.

[112] Sørensen, H.P, & Mortensen K.K. (2005). Advanced genetic strategies for recombinant protein expression in *Escherichia coli. J. Biotechnol., 115*, 113-128.

[113] Heitzer, M., Eckert, A., Fuhrmann, M., & Griesbeck C. (2007). Influence of codon bias on the expression of foreign genes in microalgae. *Adv. Exp. Med. Biol. 616,* 46-53.

[114] Hatfield, G.W, & Roth, D.A. (2007). Optimizing scaleup yield for protein production: Computationally Optimized DNA Assembly (CODA) and Translation Engineeringtrade mark. *Biotechnol. Annu. Rev., 13,* 27-42.

[115] Nilsson, B. & Anderson, S. (1991). Proper and improper folding of proteins in the cellular environment. *Annu. Rev. Microbiol. 45*, 607-635.

[116] Smith, D.W. (1996). Problems of translating heterologous genes in expression systems: the role of tRNA.*Biotechnol. Prog. 12*, 417-422.

[117] Kurland, C. & Gallant, J. (1996). Errors of heterologous protein expression. *Curr. Opin. Biotechnol. 7*, 489-493.

[118] Balbás P. (2001). Understanding the art of producing protein and nonprotein molecules in *Escherichia coli*. *Mol Biotechnol., 19*, 251-267.

[119] Zalucki, Y.M, & Jennings, M.P. (2007). Experimental confirmation of a key role for non-optimal codons in protein export. *Biochem. Biophys. Res. Commun., 355*, 143-148.

[120] Zalucki, Y.M., Gittins, K.L., & Jennings, M.P. (2008). Secretory signal sequence non-optimal codons are required for expression and export of beta-lactamase. *Biochem. Biophys. Res. Commun. 366*, 135-141.

[121] Young, J.C., & Andrews, D.W. (1996). The signal recognition particle receptor alpha subunit assembles co-translationally on the endoplasmic reticulum membrane during an mRNA-encoded translation pause in vitro. *EMBO J. 15*, 172-181.

[122] Huard, F.P., Deane, C.M., & Wood, G.R. (2006). Modelling sequential protein folding under kinetic control. *Bioinformatics, 22*, e203-210.

In: Protein Misfolding
Editors: C. B. O'Doherty, A.C. Byrne
ISBN: 978-1-60456-881-3

Chapter IV

PROTEIN MISFOLDING: A NEW PARADIGM FOR HFE C282Y HEREDITARY HEMOCHROMATOSIS

Matthew W. Lawless*
Department of Clinical Medicine, Trinity College Dublin
St. James Hospital, Dublin 8, Ireland

ABSTRACT

Hereditary hemochromatosis (HH) refers to a unique clinicopathologic subset of iron overload syndromes that include the disorder related to the C282Y homozygote mutation of the hemochromatosis protein HFE (the most common form of hemochomatosis). HFE C282Y is a mutant protein the forms aggregates and is subsequently retained in the Endoplasmic Reticulum (ER). The aberrant accrual of proteins can therefore lead to ER stress which may have detrimental effects on normal intracellular function contributing to liver pathology. Recent advances are shedding light on the potential role of HFE C282Y in eliciting pathogenesis via ER stress. All available data are pointing towards a new paradigm that of, HFE C282Y HH as a conformational disorder. Conformational diseases are a class of disorders associated with an underlying accrual of mutant protein and examples include conditions such as Alzheimer's, Parkinson's, Huntington's diseases and the liver disease Z alpha 1-antitrypsin deficiency. This review sets out to examine this new paradigm for HFE C282Y HH liver pathogenesis with an emphasis on the consequences of HFE C282Y protein retention in the ER. Viewing HH from the perspective of these unified conditions will enable greater insight into the molecular events underlying this disease and provides a new vista on this metabolic iron overload disease. Furthermore, the recognition of HFE C282Y HH from this perspective suggests intriguing possibilities for new and more rational approaches to the diagnosis and treatment for this liver disease.

* Correspondence concerning this article should be addressed to: Dr. Matthew Lawless, Room 1.17 Phase 3, Hepatology Research Division, Department of Clinical Medicine, Trinity College Dublin, St. James Hospital, Dublin 8, Ireland. Email: mlawles@tcd.ie and lawlessm@hotmail.com. Tel: 353-1-896 3273; Fax: 353-1-896 3503.

Introduction

Conformational diseases arise when a constituent protein undergoes a change in size or fluctuation in shape, with resultant self-association and tissue deposition. A typical end point is the formation of cellular inclusions or pericellular deposits of conformationally unstable proteins. There is a clear differentiation between the conformation diseases and most other genetic diseases that are caused by a failure of protein production [1]. The differentiating feature of conformational disease is that some of the protein is correctly folded and is released in its normal form while the majority is incorrectly folded and forms aggregates. Many such diseases are caused by mutations in secretory proteins and range from metabolic diseases such as type 2 diabetes, which is characterized by islet amyloid derived from islet amyloid polypeptide (IAPP) and increased beta-cell apoptosis, to neurological conditions such as Alzheimer's disease. In Alzheimer's disease, for instance, the accumulation in the brain of insoluble aggregates of the amyloid-β peptide, also known as senile plaques, is thought to be a key pathological event underlying the loss of brain cells [2]. Other conformational diseases include cystic fibrosis, Z A1AT deficiency, Parkinson's disease, and Huntington's disease, each of which is associated with the intracellular accumulation of misfolded proteins and ER stress.

Hereditary hemochromatosis (HH) is a disorder of iron metabolism associated with the accumulation of iron in the body. HH is an autosomal recessive disorder resulting from mutations in the *HFE* (hemochromatosis) gene, with clinical consequences that can range from cirrhosis of the liver to diabetes, heart failure, arthritis and liver cancer [3,4].

The protein encoded by this gene is similar to MHC class I-type proteins and associates with β2-microglobulin (β2M). β2M has been shown to play an important role in regulating iron absorption. Interestingly, β2M is critical to the folding of MHC-class I molecules and the selection of MHC class I molecule-associated peptides [5]. The majority of HH patients carry an *HFE* mutation, in which the cysteine at position 282 is changed to tyrosine (C282Y). This underlying genetic mutation associated with HH was discovered as recently as 1996. This mutation prevents the formation of a disulfide bond in the α3 domain, impairing the normal association of HFE with β2M. This lack of association dramatically reduces the cell surface expression of HFE, whereby the majority of the C282Y mutant protein remains as high molecular weight aggregates, fails to undergo late Golgi processing and subsequently is retained in the ER [6-9]. Interestingly, the second most common HH mutation associated with iron overload, HFE H63D, also results in aberrant ER processing, although not to the same severity as HFE C282Y. The retention of the mutant protein in the ER has been suggested to provide a possible basis for the pathogenesis of this condition [9,10].

The inclusion of HFE C282Y HH in the family of conformational diseases is based on common criteria. Studies have demonstrated the characteristic feature of limited expression of the mutant C282Y HFE protein on the cell surface [9,11,12]. Furthermore, studies have shown its retention in the ER is followed by marked degradation of protein with the potential to elicit ER stress, thereby illustrating all of the classical hallmarks of a conformational disease. Indeed, the state of the art at present strongly points to the occurrence of ER stress associated with the HFE C282Y mutant protein in various models, but a great deal of research is needed to understand the contribution that this has on other proteins involved in iron metabolism. Importantly, the HFE C282Y mutant elicited the unfolded protein response (UPR). Indeed, the UPR is currently understood to be involved in human disease [13], including having a role in diabetes mellitus, atherosclerosis, and neoplasia, all of which are complications seen in patients suffering from HFE C282Y hereditary hemochromatosis. Many questions remain to be answered in this emerging area of research in regards to HH liver disease onset and progression. A fundamental question is whether the liver copes with the C282Y HFE mutant protein by physiological UPR or whether it fails, leading to ER stress-associated clinicopathobiology.

To date, HH has been largely viewed as a metabolic iron overload disorder without much emphasis placed on the role of the C282Y mutation. Against the backdrop of recent and exciting advances, this timely review may provide a forum for the advancement of this new paradigm of ER stress in HH hepatotoxicity. In order to better control and treat this disease in a clinical setting, it is paramount to first understand the basic cell biology responses associated with HFE C282Y protein aggregation and examine its role in eliciting abnormal ER stress responses. Using the emerging evidence to support this notion, we examine HH as a conformational disease and discuss ER stress activation associated with HFE C282Y protein retention. Furthermore, we extrapolate from recent findings to generate a new paradigm for HFE C282Y-induced tissue injury. Indeed, while reviews on HH have to date largely dealt with the end-stage consequence of this disorder due to iron overload, we will emphasize and focus on upstream molecular consequences of the presence of the HFE C282Y mutant protein. Furthermore, we will discuss recent advances in the field of ER stress-induced liver disease to provide a novel mechanistic insight into the molecular pathogenicity of this liver disease.

In the following sections we will illustrate how viewing HFE C282Y HH from this perspective and applying the common concepts of a conformational disorder may therefore prove useful in furthering the understanding of both the pathogenesis and variable clinical presentation among C282Y homozygous individuals.

Protein Misfolding

A decade ago, it was discovered that a number of diseases with different pathologies are caused by protein misfolding. The basis for these diseases, including Alzheimer's, Parkinson's, Huntington's and Z A1AT deficiency [1,14], is a cellular inability to degrade misfolded or damaged proteins and the formation of cytotoxic intracellular or extracellular aggregates. Their pathology arises from the aggregation process; thus, it can be considered a gain-of-toxicity function. These types of conformational diseases are multiparametric in their causes, having both genetic and environmental contributing factors. Some are, however, predominantly genetic diseases, such as early onset Alzheimer's and Parkinson's, and some may be considered monogenic, such as Huntington's and the liver disease Z A1AT deficiency [2,13,15-18].

A common feature of almost all diseases of protein conformation is the formation of aggregates caused by destabilization of the α-helical structure coupled with the simultaneous formation of a β-sheet [1,2]. Such a linkage with a pleated strand from one molecule being inserted into a pleated sheet of the next is undemanding in both structural and energy requirements and can therefore readily take place between a variety of molecules. In Alzheimer's disease, for instance, the accumulation in the brain of insoluble aggregates of the amyloid-β peptide, also known as senile plaques, is thought to be a key pathological event underlying the loss of brain cells [19]. These aggregates tend to resist degradation and accumulate in inclusion bodies, which are usually present in low copy number, most often only one per cell retained in the ER [20-23]. The ER is one of the largest cell organelles, comprising a network of interconnected, closed membrane-bound vesicles. The ER is the site of synthesis, folding and modification of secretory and cell-surface proteins and serves many essential functions, facilitating the modification, folding and transport of proteins and synthesizing lipids and cholesterol. Only correctly folded proteins are transported out of the ER, while incompletely folded proteins are retained in the organelle to complete the folding process or to be targeted for destruction. Therefore, when ER homeostasis is altered, the ER responds by inducing the expression of specific genes in an attempt to restore normal ER stability (Figure 1). The principle mechanisms of ER stress are primarily contained within four pillars: (1) ER protein degradation (ERAD), (2) the endoplasmic overload response (EOR), (3) the unfolded protein response (UPR) and (4) the cellular death pathway.

In the following sections we will provide an overview of recent advances illustrating the formation and retention of C282Y HH protein aggregates in the ER and ideas emerging in the field of ER stress activation.

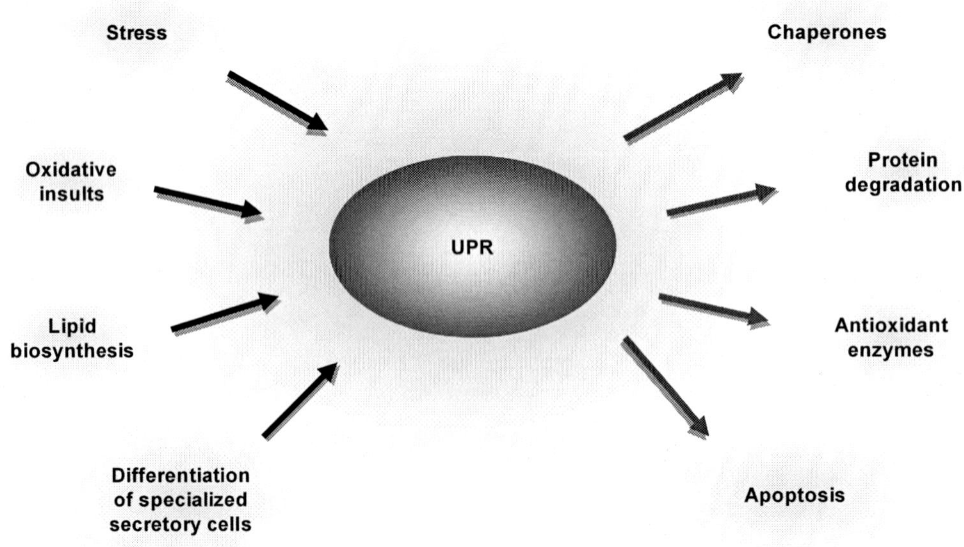

Figure 1. Inducers and responses of UPR activation and regulatory expression. Activation of UPR by stress, oxidative insults, differentiation of specialized secretory cells, signaling and lipid biosynthesis. Expression of a number of genes involved in regulation of chaperones, antioxidant enzymes and protein folding.

The Importance of ER Stress Paradigm for HFE C282Y HH

Evidence is emerging that a critical determinant in the pathogenesis of liver diseases is how the liver responses to stress. Stress can emerge from exogenous sources thorough membrane receptors or internally from any organelle, e.g. the nucleus, mitochondria, cytoskeleton, or ER. In each case, molecular signals are activated that elicit built-in mechanisms to promote a protective response or, when overwhelmed, a pathological response. Indeed, in the liver ER stress stands at the crossroads of life and death. This duality may be viewed as an example of a yin/yang relationship, perhaps nature's way to accomplish subtle regulatory changes as a response to ever-shifting physiological requirements. Evidence has been mounting for an important role of the ER stress response in the pathogenesis of chronic viral hepatitis, nonalcoholic fatty liver disease, genetic liver disorders of protein misfolding, and alcoholic liver disease. An increasing interest in the potential role of ER stress in liver disease is leading to the anticipation of a greater understanding of the pathophysiology of liver disease. Indeed, it is know known that when hepatocytes undergo sustained ER stress, end-stage apoptosis and steatosis can occur establishing a potential for tumorigenesis.

Recent *in vitro* and *in vivo* data has shown ER stress activation associated with HFE C282Y protein expression and retention in the ER. In this regard, HFE C282Y protein appears to have properties predisposing it to elicit cellular ER stress. As such, it is possible to now associate this liver disorder with the above liver diseases in terms of ER stress. These findings raise the prospect of a better understanding of the pathogenesis of HH. In order to advance this new paradigm for HFE C282Y HH liver disease, we will discuss the currently available evidence linking these concepts and extend the hypothesis that C282Y HFE HH pathogenesis is, at least partly, a consequence of ER stress.

The *HFE* Gene Structure and Function

The *HFE* gene is located on chromosome 6-band p22.1 [7]. The HFE protein is closely related to major histocompatibility complex (MHC) class I molecules. MHC class I molecules, like HFE, are also encoded on chromosome 6 and interact with β2-microglobulin [6,7,24]. However, whereas MHC molecules function in immune responses, the HFE protein has no clear immunoregulatory role to date. HFE is a transmembrane glycoprotein with both intracellular and extracellular domains [25]. Localization studies of iron-metabolism-related mRNAs in rat liver indicate that HFE expression is predominantly within hepatocytes. The HFE protein contains a binding site for β2-microglobulin. Interestingly, β2-microglobulin is critical to the folding of MHC-class 1 molecules and selection of MHC class 1 molecule-associated peptides. Both HFE and β2-microglobulin are required for iron regulation. This is illustrated by the finding that both the β2-microglobulin and *HFE* knock-out mice present with a hemochromatosis phenotype [26], [24] HFE protein interacts with and forms a complex with the transferrin receptor, TfR1. TfR1 is responsible for receptor-mediated endocytosis of iron-bound transferrin [4,27]. The interaction between HFE and TfR is proposed to be necessary for the normal control of iron absorption; mutant mice homozygous for both *TfR* and *HFE* loss accumulate more liver iron than *HFE*-only knock-out mice. In addition, HFE has also been found to independently modulate TfR2 levels in hepatic cells [28-31]. Therefore, understanding the role of HFE in normal iron regulation and abnormal iron overload will provide important molecular insights into cellular iron haemostasis.

Iron Regulation

Iron is an essential metal in the body. It is an important bio-catalyst of oxidation–reduction reactions and is indispensable for basic cellular functions [32]. The capacity of iron to readily exchange electrons in aerobic environments makes this metal vital for fundamental cell functions, such as DNA synthesis, transport of oxygen and electrons, and cellular respiration [33-35]. Plasma levels of iron are tightly regulated to ensure a daily supply of approximately 20 mg to the bone marrow [4]. In the bone marrow, more than two-thirds of the body's iron content (3-5 g in healthy adult humans) is incorporated into haemoglobin in erythroid precursors and mature red blood cells [35]. Senescent erythrocytes are continually broken down by macrophages, and this process returns iron to the plasma pool. The iron that is recycled by macrophages is loaded onto serum transferrin and returned to the bone marrow for reincorporation into new erythrocyte precursors. A small percentage of the body's iron is also present in enterocytes in the storage compound ferritin (1-2 mg). The remainder of the body's iron, approximately 1000 mg, is stored primarily in the liver, in hepatocytes As the liver is the main site of iron storage in the body, it plays an important role in the regulation of iron homeostasis [4,32-35]. The processes of iron metabolism and regulation are major components to understanding the molecular mechanisms and pathology of hereditary hemochromatosis.

The C282Y Variant of *HFE*

HFE C282Y HH is an autosomal recessive iron-overload disorder. It is associated with mutation of the *HFE* gene, which is located on the short arm of chromosome 6 [7]. In the majority of cases, the mutation in the *HFE* gene is a single-base change (845G->A) that results in the substitution of tyrosine for cysteine at position 282 of the HFE protein (C282Y). Approximately 85% of patients with this disorder are homozygous for the HFE C282Y mutation [36].

The C282Y mutation disrupts a critical disulfide bond in the α3-domain of the HFE protein. This results in a conformational change which prevents correct binding of the mutant HFE protein to β2-microglobulin. HFE binding to β2-microglobulin is required for the stabilisation, transport and expression of HFE on the cell surface and on endosomal membranes. The result of the altered interaction of mutant *HFE* with β2-microglobulin is reduced transport to and expression on the cell surface. The mutant C282Y HFE protein becomes trapped and accumulates in the endoplasmic reticulum and middle Golgi compartments and does not undergo late Golgi processing (Figure 2). As a result, the misfolded protein is more rapidly degraded which contributes to its reduced surface expression [6-8,37-40]. Furthermore, the altered HFE protein can no longer associate with TfR1, and it is suggested that this can result in defective iron metabolism, a hallmark of HH. The different presentations for this mutation suggest that HFE function depends on the presence of modifiers acting directing or indirectly with HFE (Table 1).

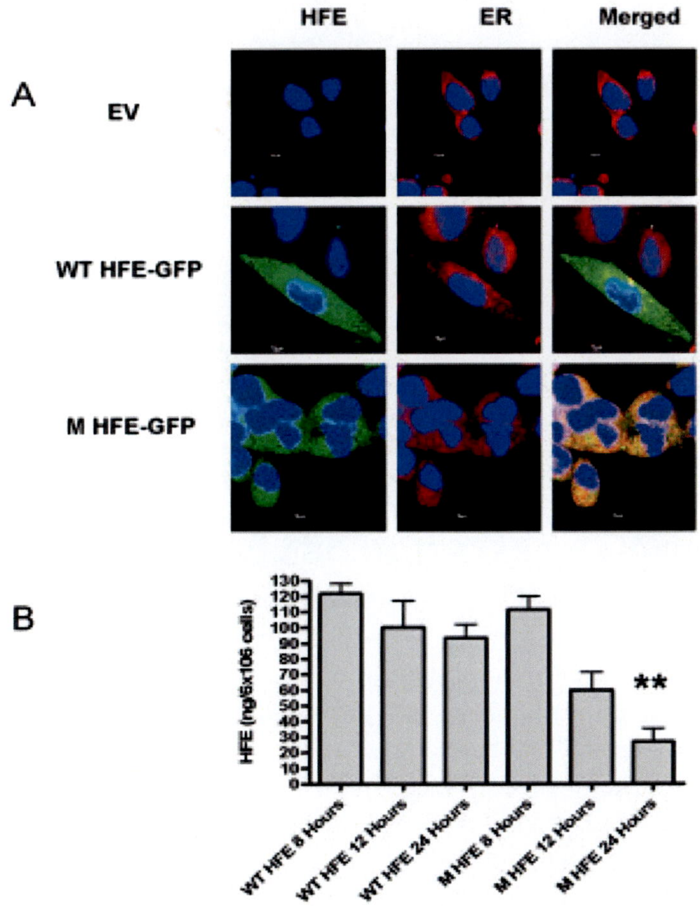

Figure 2. Intracellular localization of WT HFE and C282Y HFE in HEK293 transfected cells. A. confocal microscopy of cells cultured for 24 h after transfection with EV-GFP, WT-HFE GFP and C282Y-HFE GFP stained with an antibody for the ER-resident protein calnexin and probed with Texas Red labeled secondary antibody. In the merged images, the yellow color corresponds to areas in which the green and red stainings overlap. The nucleus appears blue due to DNA staining with 4',6-diamidino-2-phenylindole. Scale bar, 10 μm. (B) HFE expression in transfected HEK293 cell lysates. HEK 293 cells (6×10^6) were transfected with pcDNA3.1 (empty vector, EV), WT HFE, or C282Y HFE (M) expression plasmids. C282Y HFE-His tag protein production in lysates was measured by ELISA. C282Y HFE-His tag-transfected cells showed markedly accelerated degradation of the C282Y HFE protein compared to N HFE protein (**, $p = 0.0012$) at 24 hours compared to WT HFE expression. Assays were performed in triplicate. (Image previously published by Lawless *et al.*, BMC Cell Biology 2007 PMID: 17650303).

Table 1. Indicating the known or postulated gene products and functions of genes with which HFE importantly interacts

Affected Gene (symbol/Location)	Known or postulated gene product function
Hemochromatosis Gene (HFE/6q21.3)	■ Interaction with transferin receptor 1 Uptake of transferin-iron ■ Hepcidin antimicrobial peptide (HAMP/19q13.1) regulator. Down regulation of iron efflux from macropages, enterocytes, placenta ■ Transferin-receptor 2 (TFR2/7q22) Uptake of tranferrin and non tranferrin bound iron Hepcidin regulator

Clinical Manifestation of Liver Disease in HFE C282Y Hereditary Hemochromatosis

As humans have no known mechanism to control iron excretion, excess iron absorption leads to accumulation of iron in tissues. This deposition of excess iron particularly affects the parenchymal organs, notably the liver and pancreas, and can threaten cell viability [39,40]. Therefore, the maintenance of iron homeostasis is crucial to health. Individuals who are homozygous for the C282Y mutation are genetically predisposed to a chain of events that may culminate in severe multi-organ damage and failure. *HFE*-related HH is representative of adult-onset HH, which is characterised by gradual iron loading and late onset parenchymal iron deposition. In HH, iron accumulates first in the transferrin pool, which results in an elevation of serum transferrin saturation. This is followed by increasing serum ferritin levels that signify iron accumulation in the tissues. Indeed, due to its redox activity, iron may also pose a threat to the cell under conditions of iron overload due to production of reactive oxygen species (ROS) and oxidative stress [25,41,42] The clinical manifestations of the disease include liver cirrhosis, arthropathy, hypermelanotic pigmentation of the skin, diabetes mellitus, endocrine dysfunction and heart failure. The liver plays a central role in iron homeostasis. It is the major site of the body's excess iron storage, which accounts for approximately 12.5 to 25% (0.5 to 1 g) of total body iron in a normal adult male. Iron is primarily sequestered in hepatocytes in the ferritin core. Hepatocytes are also the source of Tf, the iron transport protein in blood; ceruloplasmin, a serum multicopper ferroxidase that plays a role in efflux of iron out of the liver; and hepcidin, a peptide involved in the regulation of iron absorption from the intestine. Thus, hepatocytes play a major role in iron homeostasis in the body. In addition, Kupffer cells in the liver have an important role in recycling iron from senescent red blood cells. As the liver is the primary site of iron storage in the body, it is usually the first organ to be affected in this condition. Hepatocellular carcinoma is 200 times more common in hereditary hemochromatosis patients and approximately 30-45 percent of deaths in affected individuals homozygous for the HFE C282Y mutation result from hepatocellular carcinoma. Liver fibrosis associated with iron excess evolves into cirrhosis over time, which is a risk factor for the development of hepatocellular carcinoma [32].

The progression of HH may be accelerated by environmental factors. High dietary iron intake, excessive alcohol consumption and the development of other diseases, such as viral hepatitis, all promote the phenotypic expression of the disease. Upon diagnosis of HFE C282Y HH, regular treatment by phlebotomy can remove the excess iron stores. Early diagnosis of HH is important as treatment can help to prevent irreversible organ damage and preclude the reduced life expectancy associated with cirrhosis [43,44].

Potential Prospects for HFE C282Y HH

The functional significance of HFE C282Y mutant protein is now at least partly known by the abnormality in protein trafficking and/or cell-surface expression of HFE that is suggested to play a pivotal role in the onset of HH disease. Studies have revealed that wild-type HFE binds to β_2-microglobulin and that the C282Y mutation completely abrogates this interaction and disrupts intracellular protein trafficking. The C282Y mutation results in the intracellular sequestration of the HFE protein, leading to HH disease. The C282Y mutant protein co-localizes with the ER resident protein calnexin and undergoes marked degradation as compared to WT HFE by accumulating in the ER and middle Golgi compartment, and consequently undergoing degradation (Figure 2). Recent studies using proteomic analysis of hepatic iron overload in mice have shown increased levels of GRP78/Bip protein [9].This up-regulation has previously been observed in human cells and it has been suggested that up-regulation of GRP78/Bip may indicate an increased demand for re-folding or retention of proteins in the ER of iron-overloaded cells [45,465]. Therefore, HFE C282Y may be suggested to have a dual role in pathogenesis via (1) altering HFEs normal interaction with other iron regulatory proteins along the iron hemostasis cascade and (2) predisposing the ER to stress by it having to deal with an abnormal cargo of mutant protein (Figure 3).

Both HFE and β2M are required for iron regulation and homeostasis. β_2M has been shown to play an important role in regulating iron absorption. This is illustrated by the finding that β2M knock-out mice and *HFE* knock-out mice present with a hemochromatosis phenotype. Furthermore, HFE protein interacts and forms a complex with the transferrin receptor, TfR1 responsible for receptor-mediated endocytosis of iron-bound transferring. This interaction between HFE and TfR is proposed to be necessary for the normal control of iron absorption. Indeed, transfection of HFE into HeLa and HEK293 cells has demonstrated that HFE expression lowers iron levels within these cells. This provides evidence of a role for HFE in the regulation of iron, whereby HFE decreases ferritin levels, increases TFR1 and

IRE levels, and lowers Tf-mediated iron-uptake into cells [40,42,47,48]. It was proposed that the lower iron levels originally observed in HeLa cells might be an artefact due to a lack of β2M expression in human cells transfected with HFE alone. However, more recent observations have discounted this hypothesis. Indeed, transfection of a human lung carcinoma cell line with additional β2M did not reverse the low ferritin levels and elevated TfR1 levels resulting from HFE over-expression. In addition, it has now been demonstrated that in HeLa cells, which show a reduced iron overload phenotype upon induction of HFE, β2-microglobulin levels are not the limiting factor. Furthermore, recent studies have shown that the labile iron pool is consistently decreased when HFE expression is increased, suggesting that the observed induction of HFE has a protective function by limiting cellular iron exposure during stress [49]. Other studies attribute the lower iron phenotype to a reduction in the rate of endocytosis of the HFE/TfR1 complex or a change in the ratio of Tf-mediated and non-Tf-mediated iron uptake. In contrast, HFE expression in monocytes has been shown to result in a higher iron level phenotype [50,51].

In the liver, the predominant site of HFE expression is hepatocytes. Transferrin receptor 2 (TfR2), a homolog of TfR1, is also primarily expressed in hepatocytes and involved in the regulation of the iron pool. Moreover, mutations in TfR2 results in HH type 3, a rare form of iron overload. Recent data reveals that HFE modulates TfR2 levels in hepatic cells in an independent fashion. Furthermore, lack of functional HFE is suggested to lead to a muted hepicidin response in mouse models of HH. Data regarding a role for hepcidin in HH iron loading, however, has been inconsistent. Nonetheless, these studies provide strong evidence for the role of HFE sequestration in iron regulation.

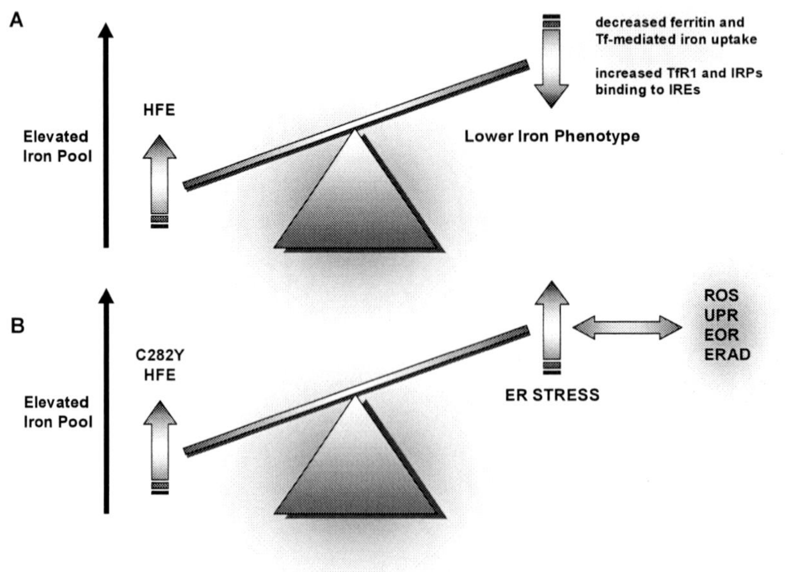

Figure 3. Proposed model of C282Y HFE toxicity from the point of view of protein misfolding. A. Under elevated levels of iron in the body normal HFE expression is induced in an effort to relieve the iron overload burden via HFE interactions with key iron regulatory molecules. B. Indicating how the presence of the C282Y HFE mutation alters HFE normal response to iron overload burden resulting in the potential occurrence of the abnormal ER stress responses.

Abnormalities in the expression of HFE C282Y protein resulting in abnormal ER retention and degradation would likely alter HFE interactions with iron regulating proteins (e.g., β2M, TfR1 and TfR2) to alter iron haemostasis (Figure 3) [9,11,49]. Indeed, this mechanism could be the basis for C282Y HFE HH pathogenicity whereby as iron levels rise, the HFE protein is also upregulated in an attempt to decrease the iron labile pool in the body. As iron levels rise, the synthesis and sequestration of HFE is elevated; thus, the client load of mutant HFE C282Y protein would exacerbate the cargo load within the ER and act as an ER stress stimuli. Mutant proteins with such impaired functions have been shown to cause pathological ER stress in a variety of disorders *via* activation of several pathways, such as EOR, UPR activation and cellular death [13,52-55]. Indeed, Lee *et al.* found that increased HFE expression correlated with ER stress and could be induced by serum deprivation and β-amyloid levels [49]. While the C282Y mutation results in an altered disease state, phenotype studies in mice have shown that it does not completely disrupt the function of HFE, emphasizing the importance of additional insults. In this regard, activation of the UPR pathway could have an initial protective role. Therefore, this might explain the milder phenotype of C282Y HFE mice as compared to the similar but more severe phenotype of the HFE knock-out mice [56]. It is possible, however, that differences in the quality control systems within the ER can explain some of the phenotypic variation observed in HFE C282Y HH patients.

Indeed, evidence now exists that aberrant ER stress is a consequence of HFE C282Y protein expression. Studies have demonstrated both *in vitro* and *in vivo* activation of ER stress as indicated by the activation of both the classical UPR and EOR pathways in association with HFE C282Y protein expression. These reports demonstrate that HFE C282Y is retained in the ER and middle Golgi compartment, and thus fails to proceed through the standard secretory pathway toward the cell membrane, resulting in ER stress and the formation of aggregates [9]. In addition, significant differences in BiP, GRP94, and CHOP expression levels that were observed between HFE wt- and C282Y 293T-expressing cells, and in PBMCs isolated from control (HFE wt) or HH patients (HFE C282Y homozygous), confirmed that the UPR is activated in response to the expression and accumulation of mutant HFE protein. In addition, these responses correlated with the degradation of HFE C282Y protein levels and the formation of HFE C282Y intracellular aggregates (Figure 4). Protein misfolding and aggregation are known to be associated with several other diseases and are particularly common in neurodegenerative diseases such as Parkinson's disease, Alzheimer's disease, or Huntington's disease. Although the identity of the aggregates and the mechanisms by which they disable and eventually kill neurons are unknown, compelling evidence strongly suggests that these aggregates may be toxic. In this regard, recent *in vitro* work revealing HFE C282Y constitutively activated both the UPR and EOR pathways through activation of NFκB coupled with the toxic cellular damage as demonstrated by end-stage apoptotic responses such as cytochrome *c* release and caspase-3 activation. This study suggests HFE C282Y has the potential to elicit end-stage ER stress responses (i.e., cellular death). These responses are associated with exacerbation of liver diseases whereby hepatic failure takes place when the amount of functioning hepatocytes decreases until the organ is not capable of fulfilling both its metabolic and synthetic functions. This excessive apoptosis lends itself to increased turnover of hepatocytes, the resultant effect of which is an exponentially increasing rate of liver scarring leading to the evolution of liver disease.

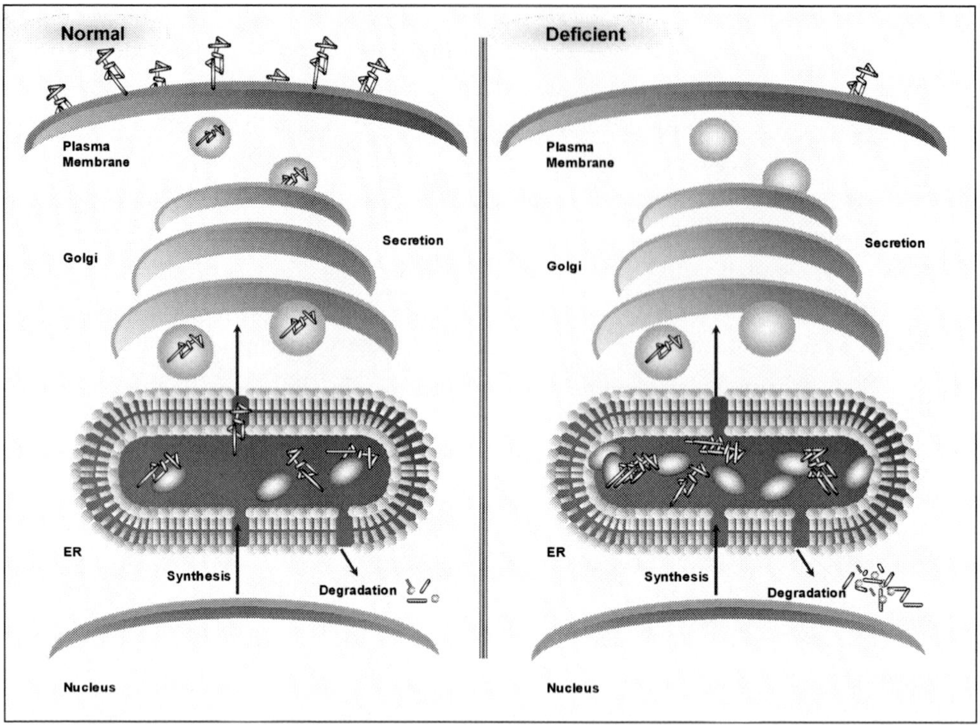

Figure 4. A conceptual model of mutant HFE C282Y protein retention in the endoplasmic reticular (ER) of a hepatocyte. Some molecules do not fold correctly and are directed to proteolytic degradation pathways with very few molecules achieving a secretion competent conformation. The increased accumulation of the HFE C282Y mutant molecules within the ER alters the ER architecture from the normal, leading to ER stress. This is manifested by the EOR and UPR response culminating in cell death and liver cell injury. However, how these signaling events sequentially interact leading to varying degrees of risk for onset of liver disease remains unclear.

Therefore, in HH, the presence of mutant HFE protein may lead to activation of CHOP through the UPR cascade, which ultimately leads to increased apoptosis. The consequences of upregulated cell death due to the C282Y *HFE* mutation in HH could explain many of the clinical features associated with this condition; for example, diminished liver function due to hepatocyte apoptosis, diabetes mellitus due to β-cell apoptosis, and heart failure due to cardiomyocyte apoptosis. Studies have shown that as HH progresses, the hepatocyte apoptotic rate increases as a function of the degree of iron storage [13]. Due to the profound physiological outcomes associated with increased apoptosis, CHOP could therefore represent two important proteins for further investigation in HH and also potential novel therapeutic targets for this disease. Downstream components of the UPR pathway include apoptotic mediators such as CHOP. CHOP is a transcription factor that is ubiquitously expressed at very low levels. In response to perturbations in ER function, CHOP expression greatly increases and this has been shown to promote growth arrest and apoptosis. However, most interestingly, CHOP expression appears to be constitutively activated by over-expression of the HH-associated C282Y *HFE* mutant, whereas in other liver diseases, such as α1-antitrypsin deficiency, the detection of CHOP requires the additional insult of a synthetic chemical ER stress-inducer [55,57,58]. This may suggest that, as compared with other liver diseases such as α1-antitrypsin deficiency, CHOP may play a more important role in the pathology associated with HH. Indeed, CHOP activation in association with HFE C282Y may prove important, indicating that the cells have a built-in ability to recognise HFE C282Y monomers and can respond by signalling the initiation of cellular destruction of the mutant proteins, which would be a likely advantage for this disease.

Pharmacological Intervention in the ER Stress Response

To date, the potential therapeutics for conformational diseases such as Z A1AT deficiency include therapies directed at preventing polymerization of the aberrant protein, chemical chaperones, and anti-inflammatory agents [59]. Polymerization may be prevented by the development of small peptide inhibitors, as shown by recent *in vitro* work in which Z A1AT polymerization was prevented by a six-mer peptide that selectively anneals to the pathogenic serpin conformation [1]. Furthermore, chemical chaperones have been tested *in vivo*, including 4-PBA. It has been suggested that the effects of chemical chaperones, in particular PBA, satisfy many of the criteria required for potential chemoprophylaxis for liver injury [60,61]. Modulation of NF-κB may also have therapeutic potential through inhibition of inflammation [55]. Tauroursodeoxycholic acid (TUDCA) is a unique bile acid that can be taken up by hepatocytes via Na^+-taurocholate cotransporting polypeptide and organic anion transporting polypeptides B1 and B3 and acts as a strong anti-apoptotic agent. It has effects that enhance its appeal as a therapeutic for liver diseases, particularly those characterized by ER stress [52,62]. For example, it has the ability to protect rat hepatocytes from calcium-dependent and ER stress-induced apoptosis. Indeed, recent studies have shown that TUDCA can inhibit caspase-3/7 and caspase-4 activation induced by thapsigargin and accumulation of aberrantly folded A1AT [13,52]. Furthermore, TUDCA may also have therapeutic potential for Z A1AT-induced liver disease as an ER stress inhibitor rather than an inhibitor of apoptosis, as it can reduce ER stress in insulin resistance and type 2 diabetes by acting as a chaperone to control ER stress and inflammation. In addition, administration of cyclosporin A to PiZ mice has been shown to be associated with a reduction in mitochondrial autophagy and injury and reduced mortality during experimental stress, whereas inhibitors of the proteosome can increase delivery of Z A1AT to the extracellular milieu [59,61-63]

Conclusion

In the past few years, we have seen remarkable advances in our understanding of how ER stress contributes to the pathology of chronic liver disease, and these advances have helped to formulate a unified view of the pathology. With the discovery of HFE C282Y-associated ER stress, the analysis of its impact on other cellular processes (namely, those involved in iron homeostasis) has now emerged as a research priority in order to establish a clear parallel between ER stress and pathophysiology in HFE C282Y HH. Furthermore, the effect of iron regulatory genes such as ferroportin, ferritin or hepcidin on ER stress signaling pathways should be further explored. Similarly, it is now paramount to fully understand the contribution of both HFE C282Y ER stress and intracellular HFE C282Y protein aggregates as risk factors for liver disease. Although phlebotomy offers an effective therapy for most individuals with the disease, for many individuals it is not an option (for example, those who undergo repeated blood transfusions for the treatment of their hemoglobinopathies). Therefore, the therapeutic potential of pharmacological chaperones that target ER stress, like TUDCA or 4 PBA, raise the prospect that chemical treatment for HFE C282Y could protect or diminish the need for phlebotomies in affected individuals. Advances in our knowledge of this subject will lead to a greater understanding of the aetiology of various disease states and raise the prospects of better therapeutics for this liver disease.

This new perspective may therefore serve as a platform to test the effects of the UPR on proteins involved in iron metabolism. Indeed, while C282Y HFE HH is considered a monogenic disease, its phenotypic expression varies considerably. Furthermore, understanding of the underlying intracellular signalling processes may further the knowledge surrounding HH pathophysiology, thereby leading to advances in treatment, diagnosis and reducing mortality.

Acknowledgements

The author kindly acknowledges Dr. Steven G. Gray and Dr. Suzanne Norris, Department of Clinical Medicine, Trinity College Dublin, St. James Hospital, for their help with this manusicipt.

References

[1] Carrell RW, Lomas DA. Conformational disease. *Lancet* 1997;350(9071):134-8.

[2] Bucciantini M, Giannoni E, Chiti F, Baroni F, Formigli L, Zurdo J, et al. Inherent toxicity of aggregates implies a common mechanism for protein misfolding diseases. *Nature* 2002;416(6880):507-11.

[3] Ajioka RS, Kushner JP. Hereditary hemochromatosis. *Semin Hematol* 2002;39(4):235-41.

[4] Pietrangelo A. Hereditary hemochromatosis--a new look at an old disease. *N Engl J Med* 2004;350(23):2383-97.

[5] Perarnau B, Siegrist CA, Gillet A, Vincent C, Kimura S, Lemonnier FA. Beta 2-microglobulin restriction of antigen presentation. *Nature* 1990;346(6286):751-4.

[6] Waheed A, Parkkila S, Zhou XY, Tomatsu S, Tsuchihashi Z, Feder JN, et al. Hereditary hemochromatosis: effects of C282Y and H63D mutations on association with beta2-microglobulin, intracellular processing, and cell surface expression of the HFE protein in COS-7 cells. *Proc Natl Acad Sci U S A* 1997;94(23):12384-9.

[7] Feder JN, Gnirke A, Thomas W, Tsuchihashi Z, Ruddy DA, Basava A, et al. A novel MHC class I-like gene is mutated in patients with hereditary haemochromatosis. *Nat Genet* 1996;13(4):399-408.

[8] Feder JN, Tsuchihashi Z, Irrinki A, Lee VK, Mapa FA, Morikang E, et al. The hemochromatosis founder mutation in HLA-H disrupts beta2-microglobulin interaction and cell surface expression. *J Biol Chem* 1997;272(22):14025-8.

[9] Lawless MW, Mankan AK, White M, O'Dwyer MJ, Norris S. Expression of hereditary hemochromatosis C282Y HFE protein in HEK293 cells activates specific endoplasmic reticulum stress responses. *BMC Cell Biol* 2007;8:30.

[10] Lawless MW, Mankan AK, Norris S. Hereditary hemochromatosis should be considered a conformational disorder. *Med Hypotheses* 2008;70(4):783-4.

[11] de Almeida SF, Fleming JV, Azevedo JE, Carmo-Fonseca M, de Sousa M. Stimulation of an unfolded protein response impairs MHC class I expression. *J Immunol* 2007;178(6):3612-9.

[12] Pinto JP, Ramos P, de Sousa M. Overexpression of HFE in HepG2 cells reveals differences in intracellular distribution and co-localization of wt- and mutated forms. *Blood Cells Mol Dis* 2007;39(1):75-81.

[13] Lawless MW, Mankan AK, Gray SG, Norris S. Endoplasmic reticulum stress-A double edged sword for Z alpha-1 antitrypsin deficiency hepatoxicity. *Int J Biochem Cell Biol* 2008.

[14] Gething MJ. Protein folding. The difference with prokaryotes. *Nature* 1997;388(6640):329, 331.

[15] Kozutsumi Y, Segal M, Normington K, Gething MJ, Sambrook J. The presence of malfolded proteins in the endoplasmic reticulum signals the induction of glucose-regulated proteins. *Nature* 1988;332(6163):462-4.

[16] Pahl HL, Baeuerle PA. The ER-overload response: activation of NF-kappa B. *Trends Biochem Sci* 1997;22(2):63-7.

[17] Smith MI, Deshmukh M. Endoplasmic reticulum stress-induced apoptosis requires bax for commitment and Apaf-1 for execution in primary neurons. *Cell Death* Differ 2007.

[18] Carrell RW, Jeppsson JO, Laurell CB, Brennan SO, Owen MC, Vaughan L, et al. Structure and variation of human alpha 1-antitrypsin. *Nature* 1982;298(5872):329-34.

[19] Berridge MJ. The endoplasmic reticulum: a multifunctional signaling organelle. *Cell Calcium* 2002;32(5-6):235-49.

[20] Kopito RR, Ron D. Conformational disease. *Nat Cell Biol* 2000;2(11):E207-9.

[21] Kopito RR. Aggresomes, inclusion bodies and protein aggregation. *Trends Cell Biol* 2000;10(12):524-30.

[22] Kopito RR, Sitia R. Aggresomes and Russell bodies. Symptoms of cellular indigestion? *EMBO Rep* 2000;1(3):225-31.

[23] Johnston JA, Dalton MJ, Gurney ME, Kopito RR. Formation of high molecular weight complexes of mutant Cu, Zn-superoxide dismutase in a mouse model for familial amyotrophic lateral sclerosis. *Proc Natl Acad Sci U S A* 2000;97(23):12571-6.

[24] Zhou XY, Tomatsu S, Fleming RE, Parkkila S, Waheed A, Jiang J, et al. HFE gene knockout produces mouse model of hereditary hemochromatosis. *Proc Natl Acad Sci USA* 1998;95(5):2492-7.

[25] Whittington CA, Kowdley KV. Review article: haemochromatosis. *Aliment Pharmacol Ther* 2002;16(12):1963-75.

[26] Rothenberg BE, Voland JR. beta2 knockout mice develop parenchymal iron overload: A putative role for class I genes of the major histocompatibility complex in iron metabolism. *Proc Natl Acad Sci U S A* 1996;93(4):1529-34.

[27] Makui H, Soares RJ, Jiang W, Constante M, Santos MM. Contribution of Hfe expression in macrophages to the regulation of hepatic hepcidin levels and iron loading. *Blood* 2005;106(6):2189-95.

[28] Chen J, Chloupkova M, Gao J, Chapman-Arvedson TL, Enns CA. HFE modulates transferrin receptor 2 levels in hepatoma cells via interactions that differ from transferrin receptor 1-HFE interactions. *J Biol Chem* 2007;282(51):36862-70.

[29] Zhang AS, Davies PS, Carlson HL, Enns CA. Mechanisms of HFE-induced regulation of iron homeostasis: Insights from the W81A HFE mutation. *Proc Natl Acad Sci U S A* 2003;100(16):9500-5.

[30] Chen J, Enns CA. The Cytoplasmic domain of transferrin receptor 2 dictates its stability and response to holo-transferrin in Hep3B cells. *J Biol Chem* 2007;282(9):6201-9.

[31] Roy CN, Carlson EJ, Anderson EL, Basava A, Starnes SM, Feder JN, et al. Interactions of the ectodomain of HFE with the transferrin receptor are critical for iron homeostasis in cells. *FEBS Lett* 2000;484(3):271-4.

[32] Corradini E, Ferrara F, Pietrangelo A. Iron and the liver. *Pediatr Endocrinol Rev* 2004;2 Suppl 2:245-8.

[33] Pietrangelo A. Mechanism of iron toxicity. *Adv Exp Med Biol* 2002;509:19-43.

[34] Pietrangelo A, Montosi G, Garuti C, Contri M, Giovannini F, Ceccarelli D, et al. Iron-induced oxidant stress in nonparenchymal liver cells: mitochondrial derangement and fibrosis in acutely iron-dosed gerbils and its prevention by silybin. *J Bioenerg Biomembr* 2002;34(1):67-79.

[35] Pietrangelo A. Physiology of iron transport and the hemochromatosis gene. *Am J Physiol Gastrointest Liver Physiol* 2002;282(3):G403-14.

[36] Shaheen NJ, Bacon BR, Grimm IS. Clinical characteristics of hereditary hemochromatosis patients who lack the C282Y mutation. *Hepatology* 1998;28(2):526-9.

[37] Fleming RE, Sly WS. Mechanisms of iron accumulation in hereditary hemochromatosis. *Annu Rev Physiol* 2002;64:663-80.

[38] Gross CN, Irrinki A, Feder JN, Enns CA. Co-trafficking of HFE, a nonclassical major histocompatibility complex class I protein, with the transferrin receptor implies a role in intracellular iron regulation. *J Biol Chem* 1998;273(34):22068-74.

[39] Pietrangelo A. Hereditary hemochromatosis. *Biochim Biophys Acta* 2006;1763(7):700-10.

[40] Pietrangelo A. Molecular insights into the pathogenesis of hereditary haemochromatosis. *Gut* 2006;55(4):564-8.

[41] Pietrangelo A. Iron chelation beyond transfusion iron overload. *Am J Hematol* 2007;82(12 Suppl):1142-6.

[42] Pietrangelo A. Hemochromatosis: an endocrine liver disease. *Hepatology* 2007;46(4):1291-301.

[43] Pietrangelo A, Oude Elferink R, Prieto J, Bacon BR. Genetics in liver diseases. *J Hepatol* 2007;46(6):1143-8.

[44] Pietrangelo A. The penetrance of hemochromatosis: mice to the rescue. *Gastroenterology* 2007;132(2):805-8.

[45] Petrak J, Myslivcova D, Man P, Cmejla R, Cmejlova J, Vyoral D, et al. Proteomic analysis of hepatic iron overload in mice suggests dysregulation of urea cycle, impairment of fatty acid oxidation, and changes in the methylation cycle. *Am J Physiol Gastrointest Liver Physiol* 2007;292(6):G1490-8.

[46] Petrak J, Myslivcova D, Halada P, Cmejla R, Cmejlova J, Vyoral D, et al. Iron-independent specific protein expression pattern in the liver of HFE-deficient mice. *Int J Biochem Cell Biol* 2007;39(5):1006-15.

[47] Enns CA. Possible roles of the hereditary hemochromatosis protein, HFE, in regulating cellular iron homeostasis. *Biol Res* 2006;39(1):105-11.

[48] Enns CA. Pumping iron: the strange partnership of the hemochromatosis protein, a class I MHC homolog, with the transferrin receptor. *Traffic* 2001;2(3):167-74.

[49] Lee SY, Connor JR. Regulation of Hfe by stress factors in BV-2 cells. *Neurobiol Aging* 2005;26(6):803-12.

[50] Davies PS, Zhang AS, Anderson EL, Roy CN, Lampson MA, McGraw TE, et al. Evidence for the interaction of the hereditary haemochromatosis protein, HFE, with the transferrin receptor in endocytic compartments. *Biochem J* 2003;373(Pt 1):145-53.

[51] Davies PS, Enns CA. Expression of the hereditary hemochromatosis protein HFE increases ferritin levels by inhibiting iron export in HT29 cells. *J Biol Chem* 2004;279(24):25085-92.

[52] Lawless MW, Mankan AK, Ryan AW, Norris S. Tauroursodeoxycholic acid: Relieving the pathogenesis of HFE C282Y hereditary hemochromatosis. *Hepatolog*y 2008.

[53] Li J, Lee B, Lee AS. Endoplasmic reticulum-stress induced apoptosis: Multiple pathways and activation of PUMA and NOXA by p53. *J Biol Chem* 2006.

[54] Ron D. Translational control in the endoplasmic reticulum stress response. *J Clin Invest* 2002;110(10):1383-8.

[55] Lawless MW, Greene CM, Mulgrew A, Taggart CC, O'Neill SJ, McElvaney NG. Activation of endoplasmic reticulum-specific stress responses associated with the conformational disease Z alpha 1-antitrypsin deficiency. *J Immunol* 2004;172(9):5722-6.

[56] Levy JE, Montross LK, Cohen DE, Fleming MD, Andrews NC. The C282Y mutation causing hereditary hemochromatosis does not produce a null allele. *Bloo*d 1999;94(1):9-11.

[57] Rudnick DA, Perlmutter DH. Alpha-1-antitrypsin deficiency: a new paradigm for hepatocellular carcinoma in genetic liver disease. *Hepatology* 2005;42(3):514-21.

[58] Hidvegi T, Schmidt BZ, Hale P, Perlmutter DH. Accumulation of mutant alpha1-antitrypsin Z in the endoplasmic reticulum activates caspases-4 and -12, NFkappaB, and BAP31 but not the unfolded protein response. *J Biol Chem* 2005;280(47):39002-15.

[59] Perlmutter DH, Brodsky JL, Balistreri WF, Trapnell BC. Molecular pathogenesis of alpha-1-antitrypsin deficiency-associated liver disease: a meeting review. *Hepatology* 2007;45(5):1313-23.

[60] Lindblad D, Blomenkamp K, Teckman J. Alpha-1-antitrypsin mutant Z protein content in individual hepatocytes correlates with cell death in a mouse model. *Hepatology* 2007;46(4):1228-35.

[61] Teckman JH, Lindblad D. Alpha-1-antitrypsin deficiency: diagnosis, pathophysiology, and management. *Curr Gastroenterol Rep* 2006;8(1):14-20.

[62] Miller SD, Greene CM, McLean C, Lawless MW, Taggart CC, O'Neill SJ, et al. Tauroursodeoxycholic acid inhibits apoptosis induced by Z alpha-1 antitrypsin via inhibition of Bad. *Hepatology* 2007;46(2):496-503.

[63] Cohen FE, Kelly JW. Therapeutic approaches to protein-misfolding diseases. *Nature* 2003;426(6968):905-9.

In: Protein Misfolding
Editors: C. B. O'Doherty, A.C. Byrne

ISBN: 978-1-60456-881-3

Chapter V

FOLDING AND AGGREGATION FEATURES OF PROTEINS

Oxana V. Galzitskaya* and Sergiy O. Garbuzynskiy

Institute of Protein Research, Russian Academy of Sciences, Pushchino, Moscow Region, Russian Federation

ABSTRACT

In addition to "normal," native protein structure, some proteins can also form alternative, misfolded structures. During the past years, it has been shown that some diseases are connected with protein misfolding and the formation of insoluble aggregates called amyloid plaques. For some proteins which are capable to form amyloid structures, those regions which are important for amyloid formation are already experimentally outlined (they are called amyloidogenic regions). In these proteins, we predicted those residues that are important for "normal" folding (that is, which are involved into the folding nucleus of the native structure) and compared them with amyloidogenic ones. The average of the predicted Φ-values (which reflect the degree of involvement of the amino acid residue into the folding nucleus) over 12 amyloidogenic regions (of 7 globular proteins in which amyloidogenic regions are now localized experimentally) is significantly greater than the average Φ-value averaged over residues outside amyloidogenic regions. This demonstrates that amino acid residues in amyloidogenic regions in average are more included into folding nucleus than amino acid residues from non-amyloidogenic regions. In total, 8 of 12 amyloidogenic fragments are located in the folding nuclei. This is an indication that amyloidogenic regions are typically incorporated into the native structure early during its formation (not later than at the rate-limiting step of a "normal" folding process).

* Correspondence concerning this article should be addressed to: Oxana V. Galzitskaya, E-mail: ogalzit@vega.protres.ru.

INTRODUCTION

In spite of the fact that each protein has its own unique, native three-dimensional structure, some cases exist when there is another rather stable structure, called an amyloid fibril. Although native structures vary greatly from protein to protein, the structures of amyloid fibrils obtained from different proteins are fairly uniform [1,2]. The observed cross β-structure is so similar from protein to protein that even antibodies created to bind amyloid fibers formed by one protein are able to bind to amyloid fibers formed by another protein [3]. The formation of amyloid fibrils is a case of protein misfolding, in which a protein folds into cross β-structure instead of folding into its native structure. In addition to proteins that form amyloid fibrils *in vivo* in various "amyloid diseases," there are many other proteins that are not implicated in amyloid diseases but form fibrils *in vitro* [4–6]. There is no sequence homology common to all such proteins or peptides.

It is known that amyloidogenic proteins and peptides possess amyloidogenic regions that are critical for fibril formation [7]. A protein can become non-amyloidogenic after mutations in these regions or after the deletion of an amyloidogenic region, while other mutations in these regions result in increased amyloidogenic propensity. In addition, when synthesized in isolation, these regions (sometimes as small as five residues long [8]) can sometimes form amyloid fibrils identical to those formed by the whole proteins [9].

It has been shown that only a few residues may be necessary to define amyloidogenic properties [10]. The sequence STVIIE was identified as the most prone to the formation of amyloid fibrils and was used as a reference to search for amyloidogenic regions in protein chains [8]. This peptide was used to verify the original hypothesis that amyloidogenic properties are determined by a short region of a protein chain [10]. The STVIIE sequence was added to the N-terminus of the water-soluble SH3 domain of α-spectrin and caused fibril formation, while the wild-type SH3 domain of α-spectrin did not form amyloid fibrils under the same conditions. The fibril core, which was inaccessible to proteases, included short sequences that belonged to the soluble globular protein and were adjacent to the amyloidogenic region [10]. Thus, short amyloidogenic regions capable of intramolecular interactions promote self-assembly of the rest of the protein into a fibrillar aggregate. If identified reliably, such amyloidogenic regions may be used as targets to prevent the formation of amyloid fibrils.

Similar results were reported for mouse β2-microglobulin, which (in contrast to human β2-microglobulin) does not form amyloid structure. Seven amino acid residues of this protein were substituted with their counterparts from human β2-microglobulin (which does form amyloids) in the region that displayed the greatest difference in primary structure between the two proteins (region 83–89). The substitution resulted in the formation of amyloid structure *in vitro* [7]. It should be noted that the peptide involved (NHVTLSQ) formed fibrils, while a peptide that consisted of the same residues joined in an arbitrary order (QVLHTSN) did not [7]. A model of a cross-β spine decorated with the remaining parts of protein molecules was proposed on the basis of the experimental findings.

Since polypeptide chains can fold into native structures or misfold into amyloid fibrils, there is a competition between the processes of folding and misfolding. Recently it has been shown that transition states of folding and aggregation for human acylphosphatase are structurally unrelated [11], suggesting that the pathways of folding and aggregation for this protein are structurally distinct. For other proteins, partially unfolded species that resemble folding intermediates have been implicated in amyloid formation [12]. This suggests that the folding and misfolding free-energy landscapes for these proteins may be more closely related than for other proteins. The identification of an intermediate as the key amyloidogenic precursor of β2-microglobulin fibril elongation under physiological conditions provides direct evidence that the folding and aggregation landscapes for this protein, at least, initially coincide and diverge only at the level of a native-like folding intermediate that contains highly localized structural variations of the native immunoglobulin fold [12]. Thus, there is no generally accepted point of view on the coincidence of regions important for amyloid formation and regions important for "normal" folding.

During folding, a protein molecule has to overcome a free-energy barrier. The most unstable structure, which corresponds to the top of the barrier (i.e., to the transition state of the folding process), is called the folding nucleus [13,14]. Since the folding nucleus is unstable, it is not easy to investigate it experimentally. A very laborious experimental method, which is called Φ-analysis [15], is developed to determine the structure of folding nuclei (then, a related method called Ψ-analysis [16] is developed). By means of the introduction of point mutations into a protein structure, it is possible to find residues, mutations of which destabilize the transition state as much as for the native state. The Φ-value is calculated for each mutation as follows:

$$\Phi = \Delta_r[F(TS) - F(U)] \,/\, \Delta_r[F(N) - F(U)], \tag{1}$$

where $\Delta_r[F(N) - F(U)]$ is the residue r mutation-induced change of the difference in the free energy between the native state N and the unfolded state U, and $\Delta_r[F(TS) - F(U)]$ is the mutation-induced change in the difference between the transition state (TS) and the unfolded state (U) free energies.

Φ-values vary from unity (when the amino acid residue is in the folding nucleus) to zero (when the amino acid residue is not incorporated into the folding nucleus). Φ-values below zero or greater than unity are indicators of the presence of non-native contacts in the transition state; such Φ-values are rare.

Since the experimental method of Φ-analysis is very laborious and there is no sufficient experimental data about folding nuclei in amyloidogenic proteins, it is useful to examine folding nuclei theoretically. At the present time, some methods exist for this purpose [17–21]. Some of them (including our method [20,22]) are able to predict positions of folding nuclei reasonably well [20–25].

The goal of this work consists in comparison of those amino acid residues which are crucial for folding and misfolding processes of the same proteins. The understanding of peculiarities of the folding process of amyloidogenic proteins can probably shed light on why these proteins are amyloidogenic. As the experimental data on both folding nuclei and amyloidogenic regions in the same proteins are scarce, we compared experimentally found amyloidogenic regions with predicted folding nuclei for proteins with experimentally known amyloidogenic regions. We demonstrate that: most of the experimentally determined amyloidogenic fragments intersect with the regions incorporated into the predicted folding nuclei of amyloidogenic proteins. In average, Φ-values for amino acid residues in amyloidogenic regions are significantly greater than Φ-values for amino acid residues in non-amyloidogenic regions. Thus, amyloidogenic regions are important not only for the process of amyloid formation but also for the process of "normal" folding. It appears that amyloidogenic regions readily form either the folding nucleus (during "normal" folding) or amyloid fibrils (during misfolding), thus playing a crucial role in the competition between folding and misfolding. This can probably also indicate that this part of the folding nucleus can be involved in amyloid formation.

Table 1. Proteins with experimentally determined amyloidogenic regions

Name of the protein	PDB entry	Number of amino acid residues		Experimentally investigated amyloidogenic regions	*In vivo* and *in vitro* or *in vitro* only
		in the protein	in the used 3D structure (region[a])		
Acylphosphatase	1aps[b]	98	98 (1–98)	16–31 [11]; 87–98 [11]	*in vitro*
β2-microglobulin	1im9	99	99 (1–99)	20–41 [32]; 59–71 [33]; 83–89 [7]	*in vivo* and *in vitro*
Gelsolin	1kcq	104	104 (158–261)	52–62 [34]	*in vitro*
Transthyretin	1bm7	127	114 (10–123)	10–19 [35]; 105–115 [36]	*in vivo* and *in vitro*
Lysozyme	1931	130	129 (1–129)	49–64 [37]	*in vivo* and *in vitro*
Myoglobin	1wla	153	153 (1–153)	7–18 [38]; 101–118 [39]	*in vitro*
Human prion	1qm0	253	143 (125–228)	169–213 [40]	*in vivo* and *in vitro*

[a]The numeration (in brackets) corresponds to that in the used PDB entry.

[b]Experimentally, amyloidogenic regions were revealed for human acylphosphatase. 3D structure of this protein is unknown. However, there is a 3D structure for horse acylphosphatase sequence of which is highly similar to that of the human protein (sequence identity 95%).

Results and Discussion

A Description of the Database of Globular Proteins with Experimentally Determined Amyloidogenic Regions

To investigate folding/unfolding behavior of amyloidogenic proteins, we have constructed a database of globular proteins with experimentally revealed amyloidogenic regions. From literature data, we selected those globular proteins in which the position of amyloidogenic regions is known from experimental data. The database now includes seven proteins (Table 1). The size of the proteins varies from 98 (acylphosphatase) to 253 (human prion protein) amino acid residues. Although each of these proteins can form amyloid fibrils (*in vivo* or *in vitro*), the globular states of these proteins are rather different. Though some amyloidogenic proteins in their native globular states are β-structured (for example, β2-microglobulin), the other proteins are α-helical (for example, myoglobin) or possess large unstructured regions (for example, human prion protein). The number and size of the experimentally revealed amyloidogenic regions also vary (see Table 1). Altogether, the seven amyloidogenic proteins contain 12 amyloidogenic regions which cover almost a quarter of the length of the proteins (24% of amino acid residues).

Theoretical Search for Folding Nuclei of Amyloidogenic Proteins

Using a theoretical method, we searched for the position of the folding nuclei in the amyloidogenic proteins during their "normal" folding process from unfolded state into their folded native structure. Using the method which (as we have previously shown [20,22,24,25]) is able to predict the position of folding nuclei reasonably well (correlation coefficient between predicted and experimentally known Φ-values reaches 65% if the native 3D structure which is used in the modeling is of high quality, see [22]), we modeled the behavior of each protein molecule at the point of thermodynamic equilibrium (i.e. under the conditions when unfolded and native states are equally stable; under these conditions, all the other states are unstable). The behavior of the protein molecule is modeled as a simplified sequential reversible unfolding of its known native 3D structure (see Materials and Methods). The obtained network of folding/unfolding pathways is further analyzed by the dynamic programming method [26] to find transition states and (correspondingly) folding nuclei. The involvement of a residue in the folding nucleus (or rather, into the ensemble of the folding nuclei) is reflected in its Φ-value, which is unity when this residue has all of its native contacts in the transition state and zero when this residue has no contacts in the transition state. The analysis of the obtained predictions have shown that 43% of amino acid residues of the proteins have Φ-values larger than 0.5 (more than 50% of native contacts are formed) and can be attributed to folding nucleus (see Table 2). It should be mentioned that such a large size of the folding nucleus is assumed to be typical for the mid-transition point (our modeling was done just in these conditions) while far from the equilibrium (that is, in pure water etc.) the folding nucleus should be typically smaller [27]; this assumption is confirmed experimentally (when several sets of Φ-values corresponding to different conditions are published: for instance, in [28]).

Table 2. Experimentally found amyloidogenic regions and predicted folding nuclei in amyloidogenic proteins

Number of proteins in the database	Number of experimentally determined amyloidogenic regions	Average fraction of amyloidogenic regions in these proteins	Average fraction of residues in folding nuclei of these proteins	Number of amyloidogenic regions located in the folding nuclei
7	12	0.24	0.42	8

Intersection of Experimentally Determined Amyloidogenic Regions with the Predicted Folding Nuclei

We tested whether the theoretically found folding nuclei intersect with experimentally found amyloidogenic regions. It appears that 8 of 12 amyloidogenic regions are situated in folding nuclei where Φ-values are large (Table 2). For several proteins, the regions with the largest Φ-values coincide with the amyloidogenic regions (see Figure 2).

For amino acid residues in amyloidogenic regions, the average Φ-value is 0.58±0.02 while amino acid residues in non-amyloidogenic regions have the average Φ-value that is significantly smaller (0.43±0.01) (here and below, the shown error is the error of the average which is calculated as $\sigma/\sqrt{n}$ where σ is the standard deviation of the distribution, and n is the number of points). Thus, in amyloidogenic regions, an average amino acid residue has more than 50% of its contacts formed in the transition state (i.e., in the folding nucleus by our definition). The p-value obtained with Student's t-test (that is, the probability that the observed difference is accidental) is $2*10^{-11}$ that confirms that the difference between the average Φ-values of amino acid residues in amyloidogenic and in non-amyloidogenic regions is significant. Such a large average Φ-value for amyloidogenic regions (0.58) as well as such a low average Φ-value for non-amyloidogenic regions (0.43) was not obtained in 1000000 rounds of shuffling of predicted Φ-values between amyloidogenic and non-amyloidogenic regions. Thus, the probability to obtain such a difference by chance is very low ($<10^{-6}$). The shuffling resulted in distributions with the average Φ-values 0.46±0.02 for amyloidogenic regions and 0.46±0.01 for non-amyloidogenic regions (in average, t-test gave probability 0.48 that this average is really the same).

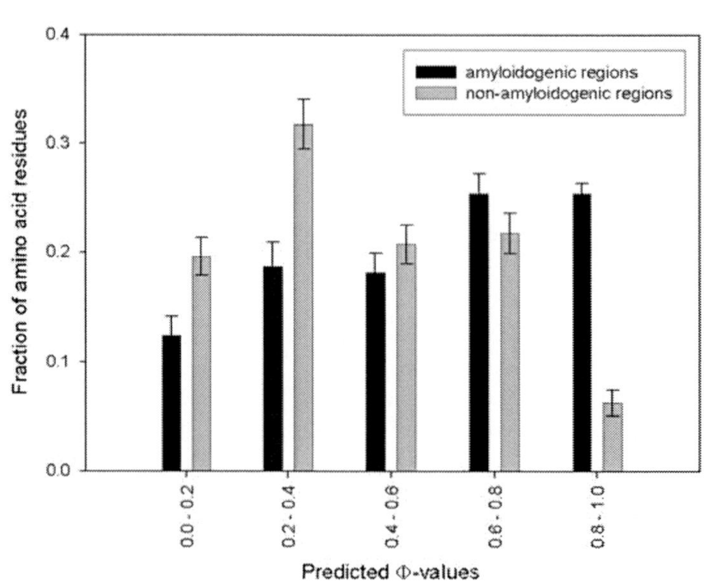

Figure 1. Distribution of predicted Φ-values in experimentally shown amyloidogenic (black bars) and non-amyloidogenic (gray bars) regions of seven investigated proteins.

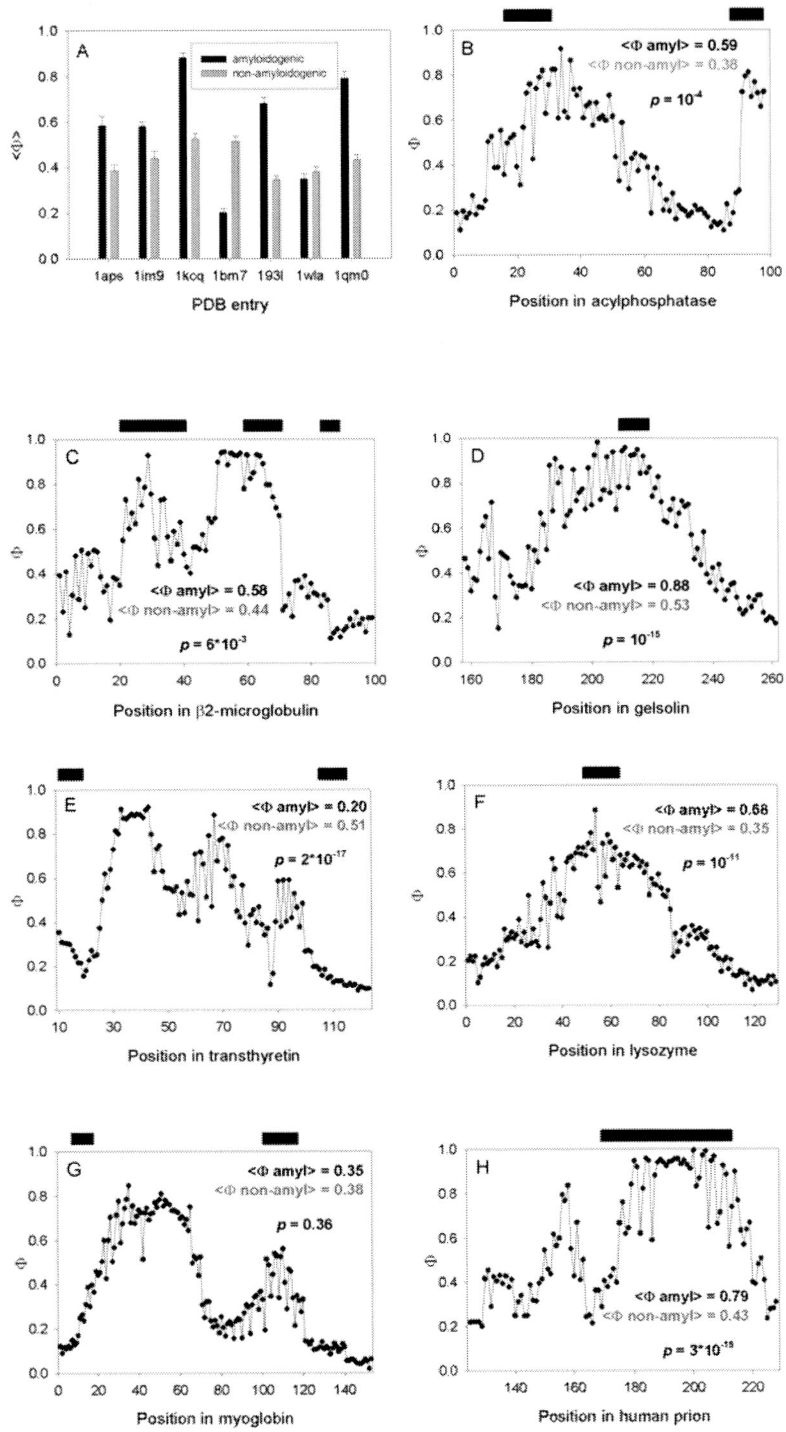

Figure 2. Φ-value profiles for the investigated amyloidogenic proteins. The positions of experimentally found amyloidogenic regions are shown as rectangles above each plot. For each protein, the average Φ-value over amyloidogenic and over non-amyloidogenic region(s) as well as probability to obtain the difference between these averages by chance (evaluated by two-tailed t-test) is given.

Shuffling of experimental amyloidogenic regions (12000 runs were done: 1000 runs for each of the 12 amyloidogenic fragments; for each amyloidogenic region, a protein was randomly chosen and then a position for the region in the protein was randomly chosen) revealed that the probability to obtain such a difference in the average Φ-values between amyloidogenic proteins and non-amyloidogenic ones as is observed in our prediction (0.15=0.58–0.43) is 0.07.

The distributions of Φ-values are shown in Figure 1. It is clear that the distributions of Φ-values in amyloidogenic and in non-amyloidogenic regions are different. It should be noted that a folding nucleus usually contains more amino acid residues (on average, 43% of amino acid residues) than are present in amyloidogenic regions (on average, 24% of amino acid residues), so the folding nucleus is typically larger (see Table 2).

Thus, we demonstrate that theoretically found folding nuclei intersect with experimentally found amyloidogenic regions. The average Φ-values are significantly greater inside amyloidogenic regions than outside them. It means that amyloidogenic regions tend to be present in the folding nucleus of a protein when the protein folds into its native structure.

The profiles of Φ-values and the positions of amyloidogenic regions for individual proteins are shown in Figure 2. Totally, amyloidogenic regions correspond to regions with large predicted Φ-values for five of the seven proteins and the difference in the average Φ-values is statistically reliable for each of these five proteins (see the average Φ-values inside amyloidogenic fragments and outside of them for each protein in its profile – as well as probability that this difference in the averages could be obtained by chance).

Thus, we have demonstrated that amyloidogenic regions are often predicted to be part of the folding nuclei in amyloidogenic proteins. Therefore, we can hypothesize that amyloidogenic regions often play a crucial role not only in amyloid fibril formation but also in the process of "normal" folding of amyloidogenic proteins into their native structure, since amyloidogenic regions compose part of the folding nucleus in these proteins.

Materials and Methods

Creation of the Database of Amyloidogenic Proteins

We collected globular proteins for which amyloidogenic regions have been experimentally determined and the three-dimensional structure of the native state is resolved and located in the Protein Data Bank (PDB) [29]. If 3D structure of a protein is resolved more than once, we chose the X-ray structure with the best resolution. A list of the selected proteins along with their PDB entries and experimentally determined amyloidogenic regions are presented in Table 1.

Theoretical Search for Folding Nuclei

We model the process of reversible unfolding of the protein "normal" (native) globular structure into the unfolded state at the point of thermodynamic equilibrium (where the free energies of the folded and the unfolded states are equal). Under these conditions, only the completely folded and completely unfolded states are observed while all other states (misfolded, partially folded etc.) are destabilized. The pathways of folding and unfolding coincide and all dead-ends are destabilized, so protein chain behavior can be approximated as a reversible folding/unfolding process.

We start from a known tertiary structure of the investigated protein in the native state and stepwise unfold its amino acid residues. A link of five residues is unfolded at a single step (this size of link was shown to be optimal for predictions of folding nuclei in globular proteins) [22]. Finally, we obtain the completely unfolded state. The unfolded residues lose all contacts that existed in the tertiary structure but gain the entropy of the coil (except for the entropy which is expended on fixation of unfolded loops that protrude from the folded part of the structure). Thus, any amino acid residue in our model can exist in one of two states: folded, where it has all of its native contacts with the other folded residues, and unfolded, where it has no contacts but has more entropy.

Considering all the ways of unfolding the protein structure, we obtain a network of protein folding/unfolding pathways. The free energy of any state is calculated as follows:

$$F(I) = \varepsilon \times n_I - T[\eta_I \times \sigma + \sum_{loops \in I} S_{loop}], \qquad (2)$$

where n_I is the number of native atom-atom contacts in the native-like part of I (n_I does not include contacts of neighbor residues, which also exist in the coil; in this work, we consider two atoms to form a contact if their centers are separated by less than 6Å, which is the value of this parameter that is optimal for folding nuclei predictions [22]); ε is the energy of one contact (all contacts are assumed to be equal in energy); η_I is the number of residues in the unfolded part of I; T is the temperature; σ is the entropy difference between the coil and the native state of a residue (we take $\sigma = 2.3R$ [30], where R is the gas constant). The sum Σ is taken over all closed unfolded loops that protrude from the native-like part of I. For a loop between fixed residues k and l, S_{loop} is calculated as follows [27]:

$$S_{loop} = -{}^{5}/_{2}R \ln|k-l| - {}^{3}/_{2}R\,(r^2_{kl} - a^2) / (2Aa|k-l|), \qquad (3)$$

where r_{kl} is the distance between the C_α atoms of residues k and l, a = 3.8Å is the distance between the neighbor C_α atoms in the chain, and A is the persistence length for a polypeptide (according to Flory [31], we take A = 20Å). The term $-{}^{5}/_{2}R \ln|k-l|$ is the most significant term in this equation; the coefficient $-{}^{5}/_{2}$ (rather than Flory's value $-{}^{3}/_{2}$) follows from the condition that a loop cannot penetrate inside the globule [27].

Calculation of Φ-Values

Since in our model all native contacts have equal energy and non-native contacts are absent, the Φ-value is a fraction of native contacts formed in the transition state compared to the native state [22]:

$$\Phi = \langle \Delta_r(n_I) \rangle_{I\#} / \Delta_r(n_0), \qquad (4)$$

where $\Delta_r(n_0)$ is number of contacts deleted by the mutation in the given amino acid residue r in the native state of the protein while $<\Delta_r(n_I)>_{I\#}$ is the number of contacts deleted by the mutation averaged over all transition states found with our method. We consider all mutations on Gly; if a residue in the wild-type protein is already Gly, we take its probability to be structured in the transition state ensemble [25].

Statistical Tests

The statistical reliability of differences between the average Φ-values (namely, the probability that the difference was obtained by chance) was evaluated using two-tailed *t*-test with unequal dispersions.

Shuffling tests were performed in two variants: shuffling of Φ-values and shuffling of amyloidogenic regions.

Shuffling of Φ-values was done as follows. Φ-values were randomly distributed between amyloidogenic and non-amyloidogenic regions (10^6 times). The average Φ-values in amyloidogenic and in non-amyloidogenic regions as well as root mean square deviations of these distributions and the probabilities evaluated by *t*-test were calculated.

Shuffling of amyloidogenic fragments was done as follows. For each amyloidogenic region, at first, a protein was randomly chosen and then a position in the chosen protein was randomly chosen. After that, we calculated average Φ-values in amyloidogenic and in non-amyloidogenic regions. This was done 1000 times for each amyloidogenic fragment (there were 12 amyloidogenic fragments and thus, 12000 rounds of shuffling were made; in the case of predicted amyloidogenic regions). Finally, the number of runs in which we obtained the difference in Φ-values between amyloidogenic regions and non-amyloidogenic regions larger than in the case of our prediction was calculated. Divided by 1000, this number is, correspondingly, the probability to obtain such a difference by chance.

Acknowledgments

The authors are grateful to D. Reifsnyder for assistance in preparation of this paper.

Funding

This work was supported by the program "Molecular and Cell Biology" of Russian Academy of Sciences, by the Russian Foundation for Basic Research (08-04-00561-a and 07-04-00388-a), by the INTAS grant (05-1000004-7747), by Science School (2791.2008.4), by "Russian Science Support Foundation" and by Howard Hughes Medical Institute (grant 55005607).

REFERENCES

[1] Rudall, K. M. (1952). The proteins of the mammalian epidermis. *Adv Protein Chem.*, *7,* 253–290.

[2] Nelson, R., Sawaya, M. R., Balbirnie, M., Madsen, A. O., Riekel, C, Grothe, R., Eisenberg, D. (2005). Structure of the cross-beta spine of amyloid-like fibrils. *Nature*, *435*, 747–749.

[3] Kayed, R., Head, E., Thompson, J. L., McIntire, T. M., Milton, S. C., Cotman, C. W., Glabe, C. G. (2003). Common structure of soluble amyloid oligomers implies common mechanism of pathogenesis. *Science*, *300*, 486–489.

[4] Guijarro, J. I., Sunde, M., Jones, J. A., Campbell, I. D., Dobson, C. M. (1998). Amyloid fibril formation by an SH3 domain. *Proc Natl Acad Sci U S A*, *95*, 4224–4228.

[5] Chiti, F., Webster, P., Taddei, N., Clark, A., Stefani, M., Ramponi, G., Dobson, C. M. (1999). Designing conditions for *in vitro* formation of amyloid protofilaments and fibrils. *Proc Natl Acad Sci U S A*, *96*, 3590–3594.

[6] Fandrich, M., Fletcher, M. A., Dobson, C. M. (2001). Amyloid fibrils from muscle myoglobin. *Nature*, *410*, 165–166.

[7] Ivanova, M. I., Sawaya, M. R., Gingery, M., Attinger, A., Eisenberg, D. (2004). An amyloid-forming segment of beta2-microglobulin suggests a molecular model for the fibril. *Proc Natl Acad Sci U S A*, *101*, 10584–10589.

[8] Lopez de la Paz, M. & Serrano, L. (2004). Sequence determinants of amyloid fibril formation. *Proc Natl Acad Sci U S A*, *10*1, 87–92.

[9] Thompson, A., White, A. R., McLean, C., Masters, C. L., Cappai, R., Barrow, C. J. (2000). Amyloidogenicity and neurotoxicity of peptides corresponding to the helical regions of PrP(C). *J Neurosci Res.*, *62*, 293–301.

[10] Esteras-Chopo, A., Serrano, L., Lopez de la Paz, M. (2005). The amyloid stretch hypothesis: Recruiting proteins toward the dark side. *Proc Natl Acad Sci U S A*, *102*, 16672–16677.

[11] Chiti, F., Taddei, N., Baroni, F., Capanni, C., Stefani, M., Ramponi, G., Dobson, C. M. (2002). Kinetic partitioning of protein folding and aggregation. *Nat Struct Biol.*, *9*, 137–143.

[12] Jahn, T. R., Parker, M. J., Homans, S. W., Radford, S. E. (2006). Amyloid formation under physiological conditions proceeds *via* a native-like folding intermediate. *Nat Struct Mol Biol.*, *13*, 195–201.

[13] Fersht, A. R. (1995). Characterizing transition states in protein folding: An essential step in the puzzle. *Curr Opin Struct Biol.*, *5*, 79–84.

[14] Fersht, A. R. (1997). Nucleation mechanisms in protein folding. *Curr Opin Struct Biol.*, *7*, 3–9.

[15] Matouschek, J. T., Kellis, J., Serrano, L., Fersht, A. R. (1989). Mapping the transition state and pathway of protein folding by protein engineering. *Nature*, *340*, 122–126.

[16] Krantz, B. A., Sosnick, T. R. (2001). Engineered metal binding sites map the heterogeneous folding landscape of a coiled coil. *Nat Struct Biol.*, *8*, 1042–1047.

[17] Shakhnovich, E., Abkevich, V., Ptitsyn, O. (1996). Conserved residues and the mechanism of protein folding. *Nature*, *379*, 96–98.

[18] Daggett, V., Li, A., Itzhaki, L. S., Otzen, D. E., Fersht, A. R. (1996). Structure of the transition state for folding of a protein derived from experiment and simulation. *J Mol Biol.*, *257*, 430–440.

[19] Onuchic, J. N., Socci, N. D., Luthey-Schulten, Z., Wolynes, P. G. (1996). Protein folding funnels: The nature of the transition state ensemble. *Fold Des.*, *1*, 441–450.

[20] Galzitskaya, O. V. & Finkelstein, A. V. (1999). A theoretical search for folding/unfolding nuclei in three-dimensional protein structures. *Proc Natl Acad Sci U S A*, *96*, 11299–11304.

[21] Alm, E. & Baker, D. (1999). Prediction of protein-folding mechanisms from free-energy landscapes derived from native structures. *Proc Natl Acad Sci U S A, 96*, 11305–11310.

[22] Garbuzynskiy, S. O., Finkelstein, A. V., Galzitskaya, O. V. (2004). Outlining folding nuclei in globular proteins. *J Mol Biol.*, *336*, 509–525.

[23] Alm, E., Morozov, A. V., Kortemme, T., Baker, D. (2002). Simple physical models connect theory and experiment in protein folding kinetics. *J Mol Biol.*, *322*, 463–476.

[24] Galzitskaya, O. V., Garbuzynskiy, S. O., Finkelstein, A. V. (2005). Theoretical study of protein folding: Outlining folding nuclei and estimation of protein folding rates. *Journal of Physics: Condensed Matter*, *17*, S1539–S1551.

[25] Garbuzynskiy, S. O., Finkelstein, A. V., Galzitskaya, O. V. (2005). On the prediction of folding nuclei in globular proteins. *Mol Biol* (Moscow), *39*, 906–914.

[26] Finkelstein, A.V. & Roytberg, M. A. (1993). Computation of biopolymers: A general approach to different problems. *Biosystems*, *30*, 1–19.

[27] Finkelstein, A. V. & Badretdinov, A. Ya. (1997). Rate of protein folding near the point of thermodynamic equilibrium between the coil and the most stable chain fold. *Fold Des.*, *2*, 115–121.

[28] Ternström, T., Mayor, U., Akke, M., Oliveberg, M. (1999). From snapshot to movie: Phi analysis of protein folding transition states taken one step further. *Proc Natl Acad Sci U S A*, *96*, 14854–14859.

[29] Berman, H. M., Westbrook, J., Feng, Z., Gilliland, G., Bhat, T. N., Weissig, H., Shindyalov, I. N., Bourne, P. E. (2000). The Protein Data Bank. *Nucl Acids Res*, *28*, 235–242.

[30] Privalov, P. L. (1979). Stability of proteins: Small globular proteins. *Adv Protein Chem.*, *33*, 167–241.

[31] Flory, P. J. (1969). *Statistical mechanics of chain molecules*. New York: Interscience. 432 p.

[32] Kozhukh, G. V., Hagihara, Y., Kawakami, T., Hasegawa, K., Naiki, H., Goto, Y. (2002). Investigation of a peptide responsible for amyloid fibril formation of beta2-microglobulin by *Achromobacter* protease I. *J Biol Chem*, *277*, 1310–1315.

[33] Jones, S., Manning, J., Kad, N. M., Radford, S. E. (2003). Amyloid-forming peptides from beta2-microglobulin – Insights into the mechanism of fibril formation *in vitro*. *J Mol Biol.*, *325*, 249–257.

[34] Maury, C. P. & Nurmiaho-Lassila, E. L. (1992). Creation of amyloid fibrils from mutant Asn187 gelsolin peptides. *Biochem Biophys Res Commun., 183*, 227–231.

[35] Chamberlain, A. K., MacPhee, C. E., Zurdo, J., Morozova-Roche, L. A., Hill, H. A., Dobson, C. M., Davis, J. J. (2000). Ultrastructural organization of amyloid fibrils by atomic force microscopy. *Biophys J., 79*, 3282–3293.

[36] Jaroniec, C. P,, MacPhee, C. E., Astrof, N. S., Dobson, C. M., Griffin, R. G. (2002), Molecular conformation of a peptide fragment of transthyretin in an amyloid fibril. *Proc Natl Acad Sci U S A, 99*, 16748–16753.

[37] Krebs, M. R., Wilkins, D. K., Chung, E. W., Pitkeathly, M. C., Chamberlain, A. K., Zurdo, J., Robinson, C. V., Dobson, C. M. (2000). Formation and seeding of amyloid fibrils from wild-type hen lysozyme and a peptide fragment from the beta-domain. *J Mol Biol,. 300*, 541–549.

[38] Picotti, P., de Franceschi, G., Frare, E., Spolaore, B., Zambonin, M., Chiti, F., de Laureto, P. P., Fontana, A. (2007). Amyloid fibril formation and disaggregation of fragment 1–29 of apomyoglobin: Insights into the effect of pH on protein fibrillogenesis. *J Mol Biol., 367*, 1237–1245.

[39] Fandrich, M., Forge, V., Buder, K., Kittler, M., Dobson, C. M., Diekmann, S. (2003). Myoglobin forms amyloid fibrils by association of unfolded polypeptide segments. *Proc Natl Acad Sci U S A, 100*, 15463–15468.

[40] Lu, X., Wintrode, P. L., Surewicz, W. K. (2007). Beta-sheet core of human prion protein amyloid fibrils as determined by hydrogen/deuterium exchange. *Proc Natl Acad Sci U S A, 104*, 1510–1515.

In: Protein Misfolding
Editors: C. B. O'Doherty, A.C. Byrne

ISBN: 978-1-60456-881-3

Chapter VI

ER-Associated Folding and Degradation: Learning from Yeast?

Riccardo Bernasconi and Maurizio Molinari*
Institute for Research in Biomedicine, Via V. Vela 6, CH-6500 Bellinzona, Switzerland

Abstract

Many aspects regulating the machineries of protein folding and degradation in the yeast *S. cerevisiae* are conserved in mammalian cells. Due to the lower redundancy and to the facility of manipulating the yeast genome, many players involved in protein folding in the endoplasmic reticulum (ER) and in degradation of terminally misfolded proteins from the ER lumen (ERAD) have been first discovered in the yeast *S. cerevisiae*. *S. cerevisiae* contains simplified versions of the corresponding mammalian pathways and probably all yeast players have functional orthologs in mammalian cells. However, metazoans certainly benefit from many additional components that are still poorly characterized.

Here we highlight differences and similarities between *S. cerevisiae* and mammalian polypeptides in the exploitation of the calnexin chaperone system during glycoprotein maturation and the intervention of a recently described ER chaperone, Yos9p/OS-9, when polypeptide maturation has failed.

Introduction

* Correspondence concerning this article should be addressed to: Maurizio Molinari, e-mail: maurizio.molinari@irb.unisi.ch; Phone number: +41918200319; Fax: +41918200305.
Riccardo Bernasconi, e-mail: riccardo.bernasconi@irb.unisi.ch; Phone number: +41918200352; Fax: +41918200305

Secretory, lysosomal, endosomal, Golgi-, ER-resident and plasma membrane proteins are synthesized by ribosomes associated at the cytosolic face of the endoplasmic reticulum membrane and are co-translationally injected into the ER lumen. Most of them are N-glycosylated at Asn-Xxx-Ser/Thr sequons and acquire their native structure under the assistance of a multitude of ER-resident molecular chaperones and folding enzymes. A quality-control system surveys the ER lumen. Correctly folded and assembled proteins leave the ER and are transported through the secretory pathway at their final destination. Non-native proteins are actively retained in the folding compartment. Terminally misfolded polypeptides are retro-translocated into the cytosol where they are degraded by 26S proteasomes. Failure of protein folding programs may lead to intra- or extracellular accumulation of misfolded polypeptides. This may result in gain-of-toxic-function human diseases (e.g. liver failure and neurodegenerative syndromes). Alternatively, it may result in disposal of the folding-defective polypeptides with debilitating loss-of-function phenotypes (e.g. cystic fibrosis and hereditary lung emphysema) (Aridor, 2007). Although several key components and mechanisms regulating protein folding, quality control and degradation from the ER have been identified, we still have not precisely understood how yeast and mammalian cells fulfil these crucial tasks.

Here, we specifically discuss the different use of the calnexin chaperone system by newly synthesized glycoproteins in *S. cerevisiae* vs. mammalian cells. Additionally, we also focus on OS-9, the recently characterized mammalian ortholog of Yos9p, a protein regulating ERAD from the yeast ER.

I) Glycoprotein Folding in the ER Lumen

I. a) N-Glycosylation

Most of the proteins translocated in the ER are core glycosylated by addition of a pre-assembled oligosaccharide (glucose$_3$-mannose$_9$-N-acetylglucosamine$_2$-) (Figure 1) at Asn-Xxx-Ser/Thr sequons. Assembly of the core glycan starts at the cytosolic face of the ER and is performed by several monosaccharyltransferases, which add 2 N-acetylglucosamine (GlcNAc) and up to 5 mannose (Man) residues onto a lipid carrier (dolichol-pyrophosphate) in the ER membrane. The Man$_5$-GlcNAc$_2$- intermediate of the oligosaccharide assembly process is flipped into the ER lumen where assembly to the final form is completed by ER-resident mannosyl- and glucosyl-transferases. Only fully assembled core oligosaccharides are transferred onto nascent polypeptide chains by a multimeric enzyme, the oligosaccharyltransferase (OST). The OST scans the nascent polypeptide sequence and transfers *en bloc* the pre-assembled oligosaccharide on the side chain of asparagines in glycosylation sequons. Addition and processing of the highly hydrophilic oligosaccharides determine the fate of the associated newly synthesized polypeptides: first, N-glycans substantially increase the solubility of as yet unfolded nascent chains, thereby, inhibiting their aggregation; second, upon removal of the two outermost glucose residues (Figure 1), they recruit the lectin chaperones calnexin and calreticulin: third, if extensively de-mannosylated, they tag the associated polypeptide for translocation into the cytosol and disposal (Hebert & Molinari, 2007), as discussed in the next sections.

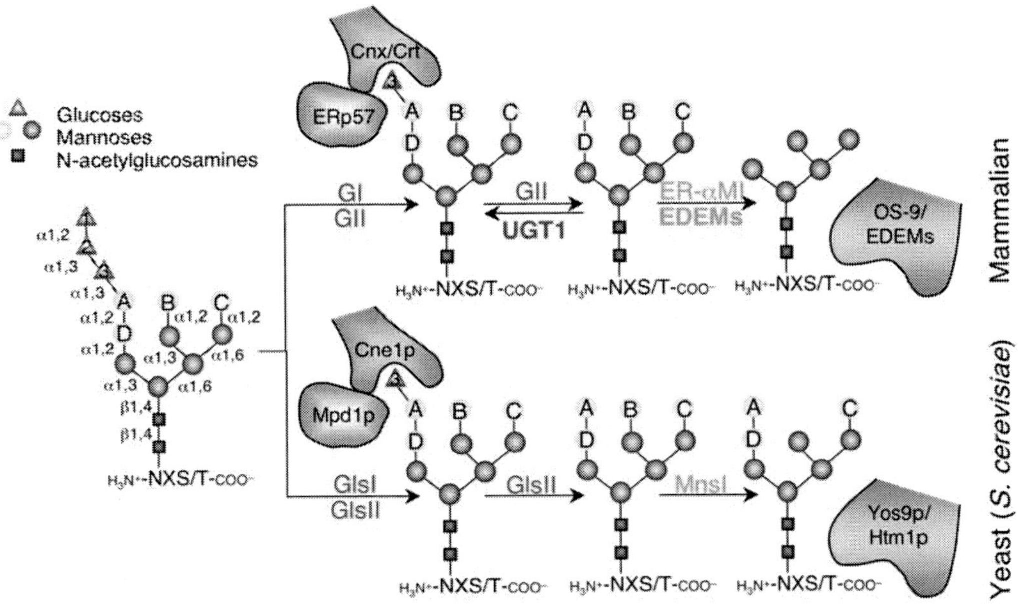

Figure 1. Schematic representation of similarities and differences in yeast and mammalian oligosaccharide processing. Cnx: calnexin; Crt: calreticulin; GI/GlsI: glucosidase I; GII/GlsII: glucosidase II; UGT1: UDP-glucose: glycoprotein glucosyltransferase; ER-αMI: ER α1,2 mannosidase I; MnsI: α mannosidase I.

I. b) The Calnexin Cycle

Immediately after addition of the core oligosaccharide onto nascent chains, the two outermost glucose residues are removed by glucosidase I and II (Figure 1). Newly synthesized glycoproteins displaying mono-glucosylated oligosaccharides bind to the lectin chaperones calnexin and calreticulin and become exposed to ERp57 (Figure 1). ERp57 is the calnexin/calreticulin-associated member of the protein disulfide isomerase superfamily, which catalyzes formation of inter- and intra-molecular disulfide bonds, a rate-limiting step of the protein folding process in the ER lumen.

Upon release from calnexin/calreticulin, native proteins are de-glucosylated by glucosidase II. They are then recruited in COPII-coated vesicles and transported at their final destination thought the secretory pathway (Gurkan, 2006). Non-native proteins released from calnexin/calreticulin are also de-glucosylated by glucosidase II. However, their non-native structure is recognized by UDP-glucose:glycoprotein glucosyltransferase (UGT1), which specifically re-glucosylates the mannose A of oligosaccharides resulting in re-association of non-native glycopolypeptides with calnexin/calreticulin for another round of folding attempts. Cycles of substrate release and re-association with calnexin and calreticulin regulated by the counteracting activities of glucosidase II and UGT1, substantially improve the efficiency of protein maturation in the mammalian ER (Figure 1) (Hebert & Molinari, 2007).

I. c) The Calnexin Cycle: Differences between the Mammalian and the Yeast System

Several mechanistical features of glycoprotein folding are conserved between mammalian and yeast. The core oligosaccharide added onto nascent chains has exactly the same composition and is actually evolutionary conserved in yeast, plants and mammals. The OST and the Asn-Xxx-Ser/Thr glycosylation sequon are also conserved. Several proteins involved in the mammalian calnexin/calreticulin chaperone system have functional orthologs in *S. cerevisiae*. For example, GlsI, GlsII and Cne1p are the glucosidase I, glucosidase II and calnexin orthologs, respectively (Helenius, 1997). The yeast oxidoreductase Mpd1p binds to Cne1p and this interaction, similarly to the calnexin-ERp57 association, enhances the catalytic activity of the oxidoreductase (Kimura, 2005). Thus, Mpd1p is a likely candidate to act in the yeast ER as a functional ortholog of ERp57. However, few components are missing in *S. cerevisiae* (e.g. calreticulin and UGT1). The absence of UGT1 introduces important mechanistic differences because it implies that in *S. cerevisiae* newly synthesized proteins cannot cycle in the calnexin chaperone system (Fernandez, 1994). Interestingly, the occurrence of UGT vary amongst various yeast strains, in fact *S. pombe* possesses UGT (Fanchiotti, 1998). Other differences are that I) Cne1p lacks the cytosolic tail that serves for calnexin association with several cytosolic and transmembrane factors; II) Cne1p is required for ERAD in lower eukaryotes (McCracken, 1996), while in mammalian cells both calreticulin and calnexin are dispensable for ERAD. Substrate entry in the mammalian calnexin/calreticulin cycle actually protects non-native proteins from premature degradation (Ayalon-Soffer, 1999; Chillaron, 2000; De Virgilio, 1999; Hebert, 1996; Kearse, 1994; Keller, 1998; Molinari, 2002; Moore, 1993).

II) Glycoprotein Misfolding in the ER Lumen

II. a) Extraction of Folding-Defective Polypeptides from the Calnexin Chaperone System: Extensive De-Mannosylation to Deviate Misfolded Proteins for ERAD

Native proteins are delivered through the secretory pathway at their final intra- or extracellular destination. Immature polypeptides must be retained in the ER lumen until achievement of their native structure. Terminally misfolded polypeptides must be translocated into the cytosol to be degraded by 26S proteasomes. Efficient execution of these three tasks is required to maintain a proper ER homeostasis (Ellgaard, 1999).

To prevent interference with maturation of incoming nascent chains, folding-defective polypeptides must eventually be extracted from the folding machinery and degraded.

In *S. cerevisiae*, the MnsI, a member of the glycosyl hydrolase family 47 (GH47) of α1,2 mannosidases, has been identified as the key enzyme that initiates protein disposal by selective removal of a single mannose residue from the N-glycan branch B (Jakob, 1998). ER α1,2 mannosidase I (ER-αMI), the mammalian ortholog, has also been proposed to act in the same way and removal of mannose B has for long time been considered the signal required for deviation of terminally misfolded proteins into the mammalian ERAD pathway (Cabral, 2001).

Careful analysis of data published in the past years and recent evidences indicate that requirements for yeast and mammalian ERAD may actually be different. Firstly, already in the early 90's, it has been shown that extensive de-mannosylation is required for mammalian ERAD (Su, 1993; Wikström, 1991). This has recently been confirmed by several studies showing that N-glycans on folding-defective polypeptides are extensively processed to $Mannose_{5\text{-}6}$ in the mammalian ER (Frenkel, 2003; Foulquier, 2002; Foulquier, 2004; Hosokawa, 2003; Kitzmuller, 2003; Lederkremer, 2005); secondly, removal of a single mannose residue also characterizes the fate of native proteins that are released from the ER into the secretory pathway (Helenius, 2001). It is therefore unlikely that this single mannose-processing event represents a strong signal for disposal. Furthermore, several studies have shown that mannose removal is still required for glycoprotein degradation in mannosyl-phosphoryldolichol-deficient cell lines (Cacan, 1992), which are characterized by addition onto nascent chains of incomplete oligosaccharides that only display removable α1,2-bonded terminal mannoses on branch A (Ermoval, 2001; Olivari, 2006). All these data imply that removal of a single mannose residue from the oligosaccharide branch B may not be sufficient to deviate a folding-defective polypeptide into the mammalian ERAD pathway. If true, this also imply that differently than the yeast ortholog, the mammalian ER-αMI can remove more than one mannose residue generating the $Man_{5\text{-}6}$ signal required in the mammalian ER.

In vitro experiments showed that at *unphysiologically* high concentrations, recombinant ER-αMI removes mannose B as well as other α1,2-bonded mannoses (Herscovics, 2002). This led to propose that this enzyme alone could be responsible for the extensive de-mannosylation of terminally misfolded polypeptides. This implies an enrichment of ER-αMI in specialized sub-regions of the mammalian ER (Lederkremer, 2005).

An additional possibility is that other α1,2-mannosidases contribute to the generation of the required ERAD signal in the mammalian ER. Likely candidates are three recently characterized ER-resident members of the GH47 family, namely EDEM1, EDEM2 and EDEM3 (Olivari, 2007). There is an ongoing debate whether the EDEM proteins are active mannosidases. Amongst data showing that they could be enzymatically active we cite that: I) all catalytic residues of α1,2-mannosidases are conserved in the EDEM proteins and the structure of ER-αMI and EDEM1 catalytic site is superimposable (Karaveg, 2005); II) up-regulation of EDEM1 and EDEM3 enhances de-mannosylation of ERAD substrates (Olivari, 2006; Hirao, 2006); III) mutation of a single catalytic residue in the putative EDEM1 and EDEM3 catalytic sites abolishes enhancement of substrate de-mannosylation (Olivari, 2006; Hirao, 2006); IV) recombinant *T. cruzi* EDEM is an active α1,2-mannosidase *in vitro* (Banerjee, 2007)

To summarize, initiation of proteins disposal in *S. cerevisiae* and in mammalian cells appears to be mechanistically different: in yeast, removal of mannose B triggers protein disposal, while in mammals extensive de-mannosylation is required (Figure 1).

Why This Difference?

To answer this question, we have to go back to the mechanisms that regulate glycoprotein-retention in the folding environment.

In *S. cerevisiae*, substrate release from Cne1p is irreversible and non-native proteins can directly be deviated into the ERAD pathway. However, in the mammalian system, the folding sensor UGT1 prolongs folding attempts in the calnexin chaperone system by persistently re-glucosylating non-native proteins eventually released from calnexin/calreticulin. Substrate cycling in the calnexin chaperone system, which is absent in *S. cerevisiae*, substantially delays onset of polypeptide disposal and can proceed as long as the folding polypeptides display the terminal a1,2-bonded mannose residue on the oligosaccharide branch A (Figure 1), the only acceptor for UGT1-mediated re-glucosylation (Hebert & Molinari, 2007). Thus, as a peculiarity of the mammalian system, removal of mannose A is required for an irreversible substrate extraction from the calnexin system, which allows deviation of terminally misfolded substrates into the ERAD pathway.

II. b) Directing ERAD Substrates to the Retro-Translocation Site

Both in yeast and in mammalian cells, the processes that regulate recognition and translocation of terminally misfolded polypeptides into the cytosol are far from being understood. Recently, many aspects and several factors involved in these processes have been characterized in *S. cerevisiae* where it has been shown that different machineries are involved in recognition, retro-translocation and polyubiquitylation of polypeptides displaying folding defects in their luminal, transmembrane or cytosolic domains. Membrane and soluble proteins with luminal defects are targeted to the ERAD-M and ERAD-L pathways, respectively (Hrd1p complex) (Carvalho, 2006; Denic, 2006). Membrane polypeptides with cytosolic lesions are targeted to the ERAD-C machinery (Doa10p complex) (Carvalho, 2006; Nakatsukasa, 2008; Vashist, 2004). ERAD-M and ERAD-C substrates are directly targeted to the retro-translocation complex, whereas ERAD-L substrates first traffic to the Golgi before returning to the ER for successive retro-translocation and polyubiquitylation (Vashist, 2004).

In *S. cerevisiae*, Yos9p belongs to the Hrd1p complex, comprising Kar2p, Hrd3p and Hrd1p, the orthologs of mammalian BiP, Sel1L and Synoviolin, respectively (Buschhorn, 2004; Bhamidipati, 2005; Carvalho, 2006; Denic, 2006; Kim, 2005; Szathmary, 2005). Yos9p is a tri-glycosylated ER-resident protein, with a C-terminus KDEL ER-retention motif. It binds terminally misfolded polypeptides and targets them to the Hrd1p complex.

Studies on the involvement of OS-9, the putative mammalian ortholog of Yos9p, in ERAD merit some discussion. It has been reported that mammalian OS-9 is a cytosolic protein, associated with the cytosolic face of the ER membrane (Litovchick, 2002). This study was followed by several other publications, in which experimental design and interpretation of the results were based on the assumption that OS-9 is a cytosolic protein

(Baek, 2005; Flashman, 2005; Wang, 2007). These studies hampered to some extent the identification of OS-9 as a crucial component of the mammalian ER quality control machinery because the cytosolic localization seemed to exclude a direct involvement of OS-9 as an ERAD regulator.

Recent studies in the lab of Ron Kopito (Christianson, 2008) and in our lab (Bernasconi, 2008) demonstrated that also in mammalian cells OS-9 is a protein of the ER lumen. These studies showed that OS-9 is a glycosylated ER-resident protein, expressed in two splice variants, OS-9.1 and OS-9.2. OS-9 variants associate with folding-defective glycoproteins, but not with native ones and have a dual function within the ER: they maintain the tightness of retention-based ER quality control by preventing forward transport of non-native conformers (this has been shown for the folding-defective $\text{Null}_{\text{hong kong}}$ variant of α1-antitrypsin (Bernasconi, 2008)) and they participate in disposal of misfolded proteins from the mammalian ER (Bernasconi, 2008; Christianson, 2008).

Our studies show that transcription of both OS-9 variants is enhanced upon activation of the Ire1/Xbp1 pathway in cells exposed to acute ER-stress. Analysis of transcriptional regulation of the mammalian orthologs of the components of the yeast Hrd1p complex reveals that all components are inducible upon ER-stress, but only OS-9 variants and Synoviolin require activation of the Ire1/Xbp1 pathway, whereas Sel1L is regulated by ATF6 (Bernasconi, 2008; Kaneko, 2007). We postulate that *retention of misfolded* versus *facilitation of disposal* functions of mammalian OS-9 variants might depend on formation of multiprotein complexes as reported for the yeast proteins, and that activation of individual stress-response pathways (Ire1-regulated versus ATF6-regulated) in specific tissues or under specific stress conditions could enhance one or the other function of OS-9.

CONCLUSION

The availability of simple techniques to manipulate the yeast genome, renders the *S. cerevisiae* a widely used model organism to characterize pathways and understand molecular processes that, in many cases, often in much more complex versions, are conserved in metazoans (Wolf, 2005). The investigation of molecular mechanisms regulating protein folding, quality control and degradation in the ER is no exception.

Here we have discussed two examples: in the first, the absence in *S. cerevisiae* of a functional ortholog of UGT1, has several implications for folding, quality control and degradation of N-glycosylated proteins, which represent the majority of the polypeptides entering the secretory pathway. By re-glucosylating proteins prematurely released from calnexin, UGT1 prolongs their retention in this glycoprotein-dedicated chaperone system and protects them from disposal. Cycling in the calnexin chaperone system can proceed as long as the non-native polypeptides display N-glycans with an intact branch A. Mannose A is, in fact, the only acceptor for the UGT1-operated re-glucosylation. Irreversible extraction from the folding environment and deviation into the ERAD pathway requires therefore extensive substrate de-mannosylation in the mammalian ER. In contrast, in *S. cerevisiae*, glycoproteins are subjected to a single round of association with Cne1p and removal of only mannose B seems to be a strong signal for polypeptide disposal.

In the second example, the different intracellular localization of the yeast Yos9p and the mammalian OS-9 seemed to indicate the involvement of the two ortholog proteins in different pathways. Thorough analysis revealed that the localization of OS-9 was actually the same in yeast and mammalian and allowed to establish a role for OS-9 in ER quality control.

References

Ayalon-Soffer, M., Shenkman, M. & Lederkremer G.Z. (1999). Differential role of mannose and glucose trimming in the ER degradation of asialoglycoprotein receptor subunits. *J Cell Sci*, *112*, 3309-3318.

Aridor, M. (2007). Visiting the ER: the endoplasmic reticulum as a target for therapeutics in traffic related diseases. *Adv Drug Deliv Rev*, *59*, 759-781.

Baek, J. H., Mahon, P. C., Oh, J., Kelly, B., Krishnamachary, B., Pearson, M., Chan, D. A., Giaccia, A. J., & Semenza, G. L. (2005). OS-9 interacts with hypoxia-inducible factor 1alpha and prolyl hydroxylases to promote oxygen-dependent degradation of HIF-1alpha. *Mol Cell*, *17*, 503-512.

Banerjee, S., Vishwanath, P., Cui, J., Kelleher, D. J., Gilmore, R., Robbins, P. W. & Samuelson, J. (2007). The evolution of N-glycan-dependent endoplasmic reticulum quality control factors for glycoprotein folding and degradation. *PNAS*, *104*,11676-11681.

Bernasconi, R., Pertel, T., Luban, J. & Molinari, M. (2008). A dual task for the Xbp1-responsive OS-9 variants in the mammalian endoplasmic reticulum: inhibiting secretion of misfolded protein conformers and enhancing their disposal. *J Biol Chem*, *283*, 16446-16454.

Bhamidipati, A., Denic, V., Quan, E.M. & Weissman, J.S. (2005). Exploration of the topological requirements of ERAD identifies Yos9p as a lectin sensor of misfolded glycoproteins in the ER lumen. *Mol Cell*, *19*, 741–751.

Buschhorn, B.A., Kostova, Z., Medicherla, B. & Wolf, D.H. (2004). A genome-wide screen identifies Yos9p as essential for ER-associated degradation of glycoproteins. *FEBS Lett*, *577*, 422–426.

Cacan, R., Villers, C., Bélard, M., Kaiden, A., Krag, S.S. & Verbert, A. (1992). Different fates of the oligosaccharide moieties of lipid intermediates. *Glycobiology*, *2*, 127-136.

Cabral, C. M., Liu, Y. & Sifers, R. N. (2001). Dissecting glycoprotein quality control in the secretory pathway. *Trends Biochem. Sci*, *26*, 619-624.

Carvalho, P., Goder, V. & Rapoport, T.A. (2006). Distinct ubiquitin-ligase complexes define convergent pathways for the degradation of ER proteins. *Cell*, *126*, 361-373.

Chillarón, J., Adán, C. & Haas, I.G. (2000). Mannosidase action, independent of glucose trimming, is essential for proteasome-mediated degradation of unassembled glycosylated Ig light chains. *Biol Chem*, *381*, 1155-1164.

Christianson, J.C., Shaler, T.A., Tyler, R.E. & Kopito, R.R. (2008). OS-9 and GRP94 deliver mutant alpha1-antitrypsin to the Hrd1-SEL1L ubiquitin ligase complex for ERAD. *Nat Cell Biol*, *10*, 272-282.

De Virgilio, M., Kitzmuller, C., Schwaiger, E., Klein, M., Kreibich, G. and Ivessa, N. E. (1999). Degradation of a short-lived glycoprotein from the lumen of the endoplasmic reticulum: the role of N-linked glycans and the unfolded protein response. *Mol Biol Cell, 10*, 4059-4073.

Denic, V., Quan, E. M. & Weissman, J. S. (2006). A luminal surveillance complex that selects misfolded glycoproteins for ER-associated degradation. *Cell, 126*, 349-359.

Ellgaard, L., Molinari, M. & Helenius, A. (1999). Setting the standards: quality control in the secretory pathway. *Science, 286*, 1882-1888.

Ermonval, M., Kitzmuller, C., Mir, A. M., Cacan, R. & Ivessa, N. E. (2001). N-glycan structure of a short-lived variant of ribophorin I expressed in the MadIA214 glycosylation-defective cell line reveals the role of a mannosidase that is not ER mannosidase I in the process of glycoprotein degradation. *Glycobiology, 11*, 565-576.

Fanchiotti, S., Fernandez, F., D'Alessio, C. & Parodi, A. J. (1998). The UDP-Glc:Glycoprotein glucosyltransferase is essential for Schizosaccharomyces pombe viability under conditions of extreme endoplasmic reticulum stress. *J Cell Biol, 143*, 625-635.

Fernández, F.S., Trombetta, S.E., Hellman, U. & Parodi, A.J. (1994). Purification to homogeneity of UDP-glucose:glycoprotein glucosyltransferase from Schizosaccharomyces pombe and apparent absence of the enzyme fro Saccharomyces cerevisiae. *J Biol Chem, 269*, 30701-30706.

Flashman, E., McDonough, M. A., & Schofield, C. J. (2005). OS-9: another piece in the HIF complex story. *Mol Cell, 17*, 472-473.

Foulquier, F., Harduin-Lepers, A., Duvet, S., Marchal, I., Mir, A. M., Delannoy, P., Chirat, F. & Cacan, R. (2002). The unfolded protein response in a dolichyl phosphate mannose-deficient Chinese hamster ovary cell line points out the key role of a demannosylation step in the quality-control mechanism of N-glycoproteins. *Biochem J, 362*, 491-498.

Foulquier, F., Duvet, S., Klein, A., Mir, A. M., Chirat, F. & Cacan, R. (2004). Endoplasmic reticulum-associated degradation of glycoproteins bearing Man5GlcNAc2 and Man9GlcNAc2 species in the MI8-5 CHO cell line. *Eur J Biochem, 271*, 398-404.

Frenkel, Z., Gregory, W., Kornfeld, S. & Lederkremer, G. Z. (2003). Endoplasmic reticulum-associated degradation of mammalian glycoproteins involves sugar chain trimming to Man6-5GlcNAc2. *J Biol Chem, 278*, 34119-34124.

Gürkan, C., Stagg, S.M., Lapointe, P. & Balch, W.E. (2006). The COPII cage: unifying principles of vesicle coat assembly. *Nat Rev Mol Cell Biol, 7*, 727-738.

Hebert, D.N., Foellmer, B. & Helenius, A. (1996). Calnexin and calreticulin promote folding, delay oligomerization and suppress degradation of influenza hemagglutinin in microsomes. *EMBO J, 15*, 2961-2968.

Hebert, D.N. & Molinari, M. (2007). In and out of the ER: protein folding, quality control, degradation, and related human diseases. *Physiol Rev, 87*, 1377-1408.

Helenius, A., Trombetta, E.S., Hebert, D.N. & Simons, J.F. (1997). Calnexin, calreticulin and the folding of glycoproteins. *Trends in Cell Biology, 7*, 193-200.

Helenius, A. & Aebi, M. (2001). Intracellular Functions of N-Linked Glycans. *Science, 291*, 2364-2369.

Helenius, A. & Aebi, M. (2004). Roles of N-linked glycans in the endoplasmic reticulum. *Annu Rev Biochem*, *73*, 1019-1049.

Herscovics, A., Romero, P. A. & Tremblay, L. O. (2002). The specificity of the yeast and human class I ER alpha 1,2-mannosidases involved in ER quality control is not as strict previously reported. *Glycobiology*, *12*, 14G-15G.

Hirao, K., Natsuka, Y., Tamura, T., Wada, I., Morito, D., Natsuka, S., Romero, P., Sleno, B., Tremblay, L. O., Herscovics, A., Nagata, K. & Hosokawa, N. (2006). EDEM3, a soluble EDEM homolog, enhances glycoprotein ERAD and mannose trimming. *J Biol Chem*, *281*, 9650-9658.

Hosokawa, N., Tremblay, L. O., You, Z., Herscovics, A., Wada, I. and Nagata, K. (2003). Enhancement of endoplasmic reticulum (ER) degradation of misfolded Null Hong Kong alpha1-antitrypsin by human ER mannosidase I. *J Biol Chem*, *278*, 26287-26294.

Jakob, C.A., Burda, P., Roth, J. & Aebi, M. (1998). Degradation of misfolded endoplasmic reticulum glycoproteins in Saccharomyces cerevisiae is determined by a specific oligosaccharide structure. *J Cell Biol*, *142*, 1223-1233.

Kaneko, M., Yasui, S., Niinuma, Y., Arai, K., Omura, T., Okuma, Y. & Nomura, Y. (2007). A different pathway in the endoplasmic reticulum stress-induced expression of human HRD1 and SEL1 genes. *FEBS Lett*, *581*, 5355-5360.

Karaveg, K., Siriwardena, A., Tempel, W., Liu, Z. J., Glushka, J., Wang, B. C. & Moremen, K. W. (2005). Mechanism of class 1 (glycosylhydrolase family 47) {alpha}-mannosidases involved in N-glycan processing and endoplasmic reticulum quality control. *J Biol Chem*, *280*, 16197-16207.

Kearse, K.P., Williams, D.B. & Singer, A. (1994). Persistence of glucose residues on core oligosaccharides prevents association of TCR alpha and TCR beta proteins with calnexin and results specifically in accelerated degradation of nascent TCR alpha proteins within the endoplasmic reticulum. *EMBO J*, *15*, 3678-3686.

Keller, S.H., Lindstrom, J. & Taylor, P. (1998). Inhibition of glucose trimming with castanospermine reduces calnexin association and promotes proteasome degradation of the alpha-subunit of the nicotinic acetylcholine receptor. *J Biol Chem*, *273*, 17064-17072.

Kim, W., Spear, E.D. & Ng, D.T. (2005). Yos9p detects and targets misfolded glycoproteins for ER-associated degradation. *Mol Cell*, *19*, 753–764.

Kimura, T., Hosoda, Y., Sato, Y., Kitamura, Y., Ikeda, T., Horibe, T. & Kikuchi, M. (2005). Interactions among yeast protein-disulfide isomerase proteins and endoplasmic reticulum chaperone proteins influence their activities. *J Biol Chem*, *280*, 31438-31441.

Kitzmuller, C., Caprini, A., Moore, S. E., Frenoy, J. P., Schwaiger, E., Kellermann, O., Ivessa, N. E. & Ermonval, M. (2003). Processing of N-linked glycans during endoplasmic-reticulum-associated degradation of a short-lived variant of ribophorin I. *Biochem J*, *376*, 687-696.

Lederkremer, G. Z. & Glickman, M. H. (2005). A window of opportunity: timing protein degradation by trimming of sugars and ubiquitins. *Trends Biochem Sci*, *30*, 297-303.

Litovchick, L., Friedmann, E., & Shaltiel, S. (2002). A selective interaction between OS-9 and the carboxyl-terminal tail of meprin beta. *J Biol Chem*, *277*, 34413-34423.

McCracken, A.A. & Brodsky J.L. (1996). Assembly of ER-associated protein degradation in vitro: dependence on cytosol, calnexin, and ATP. *J Cell Biol*, *132*, 291-298.

Molinari, M., Galli, C., Piccaluga, V., Pieren, M. and Paganetti, P. (2002). Sequential assistance of molecular chaperones and transient formation of covalent complexes during protein degradation from the ER. *J Cell Biol*, *158*, 247-257.

Moore, S.E. & Spiro, R.G. (1993). Inhibition of glucose trimming by castanospermine results in rapid degradation of unassembled major histocompatibility complex class I molecules. *J Biol Chem*, *268*, 3809-3812.

Nakatsukasa, K., Huyer, G., Michaelis, S. & Brodsky, J.L. (2008) Dissecting the ER-associated degradation of a misfolded polytopic membrane protein. *Cell*, *132*, 101-112.

Olivari, S., Cali, T., Salo, K. E., Paganetti, P., Ruddock, L. W. and Molinari, M. (2006). EDEM1 regulates ER-associated degradation by accelerating demannosylation of folding-defective polypeptides and by inhibiting their covalent aggregation. *Biochem Biophys Res Commun*, *349*, 1278-1284.

Olivari, S. & Molinari, M. (2007). Glycoprotein folding and the role of EDEM1, EDEM2 and EDEM3 in degradation of folding-defective glycoproteins. *FEBS Lett*, *581*, 3658-3664.

Su, K., Stoller, T., Rocco, J., Zemsky, J. & Green, R. (1993). Pre-Golgi degradation of yeast prepro-alpha-factor expressed in a mammalian cell. Influence of cell type-specific oligosaccharide processing on intracellular fate. *J Biol Chem*, *268*, 14301-14309.

Szathmary, R., Bielmann, R., Nita-Lazar, M., Burda, P. & Jakob C.A. (2005). Yos9 protein is essential for degradation of misfolded glycoproteins and may function as lectin in ERAD. *Mol Cell*, *19*, 765–775.

Vashist, S. & Ng, D. T. (2004). Misfolded proteins are sorted by a sequential checkpoint mechanism of ER quality control. *J Cell Biol*, *165*, 41-52.

Wang, Y., Fu, X.; Gaiser, S., Kottgen, M., Kramer-Zucker, A., Walz, G. & Wegierski, T. (2007). OS-9 Regulates the Transit and Polyubiquitination of TRPV4 in the Endoplasmic Reticulum. *J Biol Chem*, *282*, 36561-36570.

Wikström, L. & Lodish, H.F. (1991). Nonlysosomal, pre-Golgi degradation of unassembled asialoglycoprotein receptor subunits: a TLCK- and TPCK-sensitive cleavage within the ER. *J Cell Biol*, *113*, 997-1007.

Wolf, D. H. & Schafer, A. (2005). CPY* and the power of yeast genetics in the elucidation of quality control and associated protein degradation of the endoplasmic reticulum. *Curr Top Microbiol Immunol*, *300*, 41-56.

Reviewed by Omar Vanoni and Tito Calì, Institute for Research in Biomedicine, Bellinzona.

In: Protein Misfolding
Editors: C. B. O'Doherty, A.C. Byrne

ISBN: 978-1-60456-881-3

Chapter VII

From ER Stress Response to Organelle Autoregulation

Hiderou Yoshida*

Department of Biophysics, Graduate School of Science, Kyoto University, Japan

Abstract

The endoplasmic reticulum (ER) stress response has been recognized as a cytoprotective mechanism to cope with unfolded proteins accumulated in the ER, although the exact significance of the ER stress response in biological processes remains elusive. In this chapter, I will try to redefine the ER stress response as an autoregulatory mechanism that controls the amount of the ER in accordance with cellular demand. Moreover, since recent discoveries suggest that mechanisms regulating the amount or capacity of other organelles including the mitochondria, peroxisomes and Golgi apparatus also exist in eukaryotic cells, I propose the working hypothesis that the amount of eukaryotic organelles is automatically regulated by signaling between the nucleus and each organelle, to adjust each organelle's capacity according to cellular demand.

ER Stress Response as a Cytoprotective Mechanism against Stress

The endoplasmic reticulum (ER) is an organelle in which secretory proteins are synthesized. In the ER, nascent proteins are correctly folded by ER chaperones including BiP and calnexin (Gething, 1997; Helenius and Aebi, 2004), and malfolded or unfolded proteins

* Correspondence concerning this article should be addressed to: H. Yoshida, Department of Biophysics, Graduate School of Science, Kyoto University, Kitashirakawa-Oiwakecho, Sakyo-ku, Kyoto 606-8502, Japan. Tel: +81+75-753-4201; Fax: +81+75-753-3718; E-mail: hide@biophysics.mbox.media.kyoto-u.ac.jp; Web site: http://www.eonet.ne.jp/~biophysics-kyoto/

are disposed of by the mechanism of ER-associated degradation (ERAD) (Ruddock and Molinari, 2006; Brodsky, 2007). When the amount of unfolded proteins is increased by genetic mutations or environmental conditions that hamper protein folding (high levels of reducing chemicals that cleave disulfide bonds, pH changes, depletion of calcium ions necessary for BiP function, etc.), unfolded proteins are retained in the ER and tend to form protein aggregates that are highly toxic to cells and induce apoptosis (Oyadomari, et al., 2002; Sekine, et al., 2006). This stressful situation is called ER stress, and it is believed that a major cause of neurodegenerative diseases such as Alzheimer's and Parkinson's diseases are ER stress-induced apoptosis, which occurs since neurons are especially sensitive to protein aggregates (Takahashi, et al., 2003; Katayama, et al., 2004). To cope with protein aggregates accumulated in the ER, eukaryotic cells activate a self defense mechanism called the ER stress response or unfolded protein response (UPR) (Mori, 2003; Malhotra and Kaufman, 2007; Ron and Walter, 2007; Yoshida, 2007).

In mammals, the ER stress response consists of three main response pathways, that is, the ATF6, IRE1 and PERK pathways. ATF6 is a sensor molecule for ER stress in the ATF6 pathway, and an ER-localized transmembrane protein (Yoshida, et al., 1998). Upon accumulation of unfolded proteins in the ER, ATF6 is transported to the Golgi apparatus by COP I-mediated vesicular transport, and sequentially cleaved by Golgi-resident proteases called S1P and S2P, resulting in release of the cytosolic portion of ATF6, which includes both the DNA-binding and transcriptional activation domains (Haze, et al., 1999; Ye, et al., 2000; Haze, et al., 2001). In the nucleus, ATF6 binds to an enhancer called ERSE and activates transcription of ER chaperone genes (Yoshida, et al., 2001b; Yamamoto, et al., 2007). In contrast, the sensor molecule of the IRE1 pathway is IRE1, which resides in the ER and whose cytosolic portion contains an RNase domain (Tirasophon, et al., 1998; Urano, et al., 2000; Iwawaki, et al., 2001; Tan, et al., 2006). In response to ER stress, IRE1 is activated and converts the pre-mRNA of XBP1 to mature mRNA by unconventional mRNA splicing, leading to production of an active transcription factor, pXBP1(S) (Yoshida, et al., 2001a; Calfon, et al., 2002). pXBP1(S) activates transcription of ERAD components including EDEM, HRD1 and Derlin2 by forming a heterodimer with ATF6 or a homodimer (Lee, et al., 2003; Yoshida, et al., 2003). PERK is a sensor of the PERK pathway and a transmembrane kinase residing in the ER (Harding, et al., 1999). When activated by ER stress, PERK phosphorylates the a subunit of eukaryotic translational initiation factor 2 (eIF2α) and attenuates translation, thereby transiently relieving the burden on the folding machinery of the ER (Harding, et al., 2000b). Interestingly, global attenuation of translation increases translation of a transcription factor, ATF4 (Harding, et al., 2000a). ATF4 activates transcription of translation-related genes as well as genes involved in anti-oxidation (Harding, et al., 2003). Thus, eukaryotic cells have evolved highly complicated cytoprotective mechanisms to cope with ER stress.

ER Stress Response as an Organelle Autoregulatory System

To induce ER stress in vitro, cells are usually treated with chemicals such as tunicamycin (N-linked glycosylation blocker), thapsigargin (Ca-ATPase inhibitor) or dithiothreitol (a reducing agent), or forced to express mutant secretory proteins that unfold in the ER, such as the NHK variant of α1-antitrypsin. However, organisms, including humans, seldom drink such chemicals or suffer from such genetic mutations, suggesting that it is not reasonable to think that eukaryotic cells developed the ER stress response system during evolution in order to cope with such specific and rare situations. Then, what is the biological role of the ER stress response in the organism?

The first indication of the biological significance of the ER stress response came from the analysis of XBP1. It had been thought that XBP1 is a transcription factor crucial for the differentiation of mature activated B cells to antibody-secreting plasma cells, since XBP1-deficient lymphocytes created using the RAG-2 complementation system displayed a severe defect in the differentiation of plasma cells from mature B cells (Reimold, et al., 2001). In addition, since XBP1 transcripts were rapidly increased by stimuli that induce plasma cell differentiation, it had been speculated that the differentiation signal may induce expression of XBP1 and then XBP1 activates transcription of genes involved in the differentiation into plasma cell (Reimold, et al., 1996; Reimold, et al., 2001). After the surprising discovery that XBP1 is a key transcription factor of the mammalian ER stress response (Yoshida, et al., 2001a; Calfon, et al., 2002), a novel notion emerged: the differentiation signal induces differentiation into plasma cells, which produce very large amounts of antibodies that overwhelm the capacity of ER chaperones and ERAD machinery in the ER, leading to accumulation of unfolded antibody proteins and activation of ER stress response pathways. Then, the ATF6 pathway increases transcription of XBP1 (Yoshida, et al., 2000; Gunn, et al., 2004), while the IRE1 pathway splices XBP1 pre-mRNA, resulting in marked induction of pXBP1(S), which increases the expression of ER chaperones and components of the ERAD machinery as well as the ER membrane, consequently augmenting the capacity of the ER (Iwakoshi, et al., 2003; Lee, et al., 2003; Yoshida, et al., 2003; Shaffer, et al., 2004; Sriburi, et al., 2004). Thus, the differentiation of plasma cells (initiation of antibody production) is regulated by the differentiation signal, whereas the creation of enormously well developed ER structures specialized in synthesis and secretion of immunoglobulins that are a characteristic feature of plasma cells (augmentation of the ER capacity) is controlled by the ER stress response. This notion can be applied to other secretory cells, including endocrine pancreatic cells, salivary gland cells and exocrine intestinal cells. It is highly possible that the creation of the highly developed ER structures in these various kinds of secretory cells to meet the cellular demand is collectively controlled by the ER stress response, rather than separately regulated by various differentiation signals specific to each cell type. Actually, XBP1 is indispensable for developing the ER structures of pancreatic and salivary gland acinar cells (Lee, et al., 2005). Moreover, this notion could be extrapolated to all cell types, including non-secretory cells. During the cell cycle, the volume of cells is nearly doubled, and then cells divide. Similarly, the amount of the ER is increased almost twofold, and each daughter cell contains an approximately equal amount of the ER. These facts suggest the existence of regulatory mechanisms adjusting the amount or capacity of the ER to cellular demand, and I speculate that this process is regulated by the ER stress response.

AUTOREGULATORY SYSTEMS IN OTHER ORGANELLES

The notion that the biological role of the ER stress response in the organism is to regulate the amount of the ER, rather than merely to cope with stressful situations in the ER, can be generalized to other organelles. The amount of specific organelles is known to be augmented in specific situations or cell types, suggesting that it is tightly controlled by unknown mechanisms, and possibly that each organelle communicates with the nucleus to maintain its amount in accord with cellular demand, like the ER (Figure 1). Below, I briefly describe the current understanding of the regulatory systems controlling the capacity or homeostasis of the mitochondria, peroxisomes and Golgi apparatus.

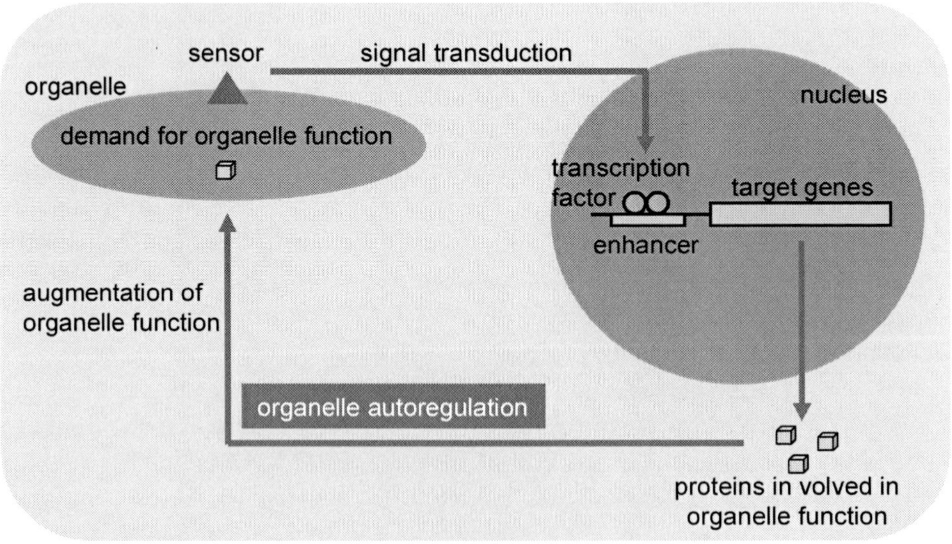

Figure 1. Working hypothesis of organelle autoregulation. When demand for organelle function is increased by differentiation, environmental conditions or cell division, a sensor monitoring a situation of the organelle is activated and transmits an emergency signal to the nucleus, leading to transcriptional induction of target genes that encode proteins involved in organelle function. This regulatory circuit may maintain the amount of each organelle in accord with cellular demand.

Mitochondria

The mitochondrion is thought to have been derived from a bacterium engulfed by a host cell, but most of the genome of the ancestral mitochondria moved to the nucleus during evolution, and the number of mitochondria in cells seems to be regulated by genes encoded by the host genome. The number of mitochondria varies among cell types. For instance, slow-twitch fibers of skeletal muscle have more mitochondria which support aerobic metabolism than fast-twitch fibers, though the molecular mechanism regulating the number of the mitochondria remains elusive. Interestingly, expression of mitochondrial chaperones, including HSP70 and HSP60, is increased in response to the accumulation of unfolded proteins in the mitochondrion by the regulatory system called the mitochondrial unfolded protein response (UPR^{mt}) (Haynes, et al., 2007; Ryan and Hoogenraad, 2007). In mammals, transcription factor CHOP is involved in transcriptional regulation of UPR^{mt} (Martinus, et al., 1996; Zhao, et al., 2002; Aldridge, et al., 2007; Horibe and Hoogenraad, 2007), whereas three regulatory factors (CLPP-1, DVE-1 and UBL-5) are involved in the regulation of UPR^{mt} in C. elegans (Yoneda, et al., 2004; Benedetti, et al., 2006; Haynes, et al., 2007). CLPP-1 is a mitochondrial matrix protease, and seems to be a sensor molecule for UPR^{mt}, whereas DVE-1 is a key transcription factor regulating the transcription of HSP60 in response to UPR^{mt}. UBL-5 is a ubiquitin-like protein, possibly involved in the signaling between the mitochondrion and the nucleus. All of these regulatory molecules are encoded in the nuclear genome, indicating that the function (and possibly capacity) of the mitochondria is regulated by nuclear-mitochondrial interactions. At present, the known targets of UPR^{mt} are limited to mitochondrial chaperones and proteases, and it is controversial whether UPR^{mt} can be defined as an autoreguratory system of the mitochondrial capacity. Further analysis of target genes of DVE-1 might reveal whether UPR^{mt} regulates the capacity of the mitochondrion as a whole, as observed in the ER stress response.

Peroxisomes

The peroxisome is an organelle that contains oxidative enzymes such as catalase and urate oxidase and carries out various oxidative reactions, such as the beta-oxidation of long chain fatty acids. The number of peroxisomes varies among cell types, and cells in the liver and kidney contain large numbers of peroxisomes (Lock, et al., 1989). The major pathway of peroxisome proliferation is not de novo synthesis or budding from the ER, but budding from pre-existing peroxisomes (Platta and Erdmann, 2007). The exact mechanism that regulates the number of peroxisomes has not been clarified, though several genes are reported to be involved in the regulation and its biogenesis (Fujiki, 2000; Fujiki, et al., 2008). Mammalian transcription factor PPARα (peroxisome proliferator-activated receptor α) is a member of the nuclear receptor superfamily and is responsible for peroxisome proliferation in response to hormones, fatty acids and chemicals such as chlofibrate (Green and Wahli, 1994). In yeast, the transcription factors Pip2p, Oaf1p and Adr1p are responsible for peroxisome induction (Rottensteiner, et al., 2003). The number of peroxisomes is also regulated by selective degradation called pexophagy that occurs via an autophagy-related process (Sakai, et al., 2006). In the methylotrophic yeast Pichia pastoris, the peroxisomes are remarkably developed for the metabolism of methanol, and when methanol-grown cells are adapted to ethanol, the peroxisomes are degraded by pexophagy (Oku, et al., 2003; Yamashita, et al., 2007). Thus, the number of peroxisomes is tightly regulated by transcriptional induction and selective degradation.

Golgi Apparatus

The Golgi apparatus is an organelle involved in the processing of secretory proteins synthesized in the ER, including modification by protein glycosylation, cleavage of proteins, protein sulfation, etc. (Berger and Roth, 1997). Sorting and transport of secretory proteins to their final destination is also an important function of the Golgi apparatus. The Golgi apparatus is extensively developed in secretory cells, such as the secretory mucous cells of Brunner's gland, whereas it is not so evident in cells secreting less protein such as fibroblasts. Interestingly, the Golgi apparatus in both acinar cells of the mammary gland and prolactin cells of the pituitary gland develops in lactating female rats, whereas its size is rapidly reduced after removal of the litter (Clermont, et al., 1993; Rambourg, et al., 1993). When pups are returned to the mother, the Golgi apparatus swiftly recovers its size in response to the sucking stimulus. These observations strongly suggest that the amount of the Golgi apparatus is strictly and dynamically regulated in accordance with cellular demand, but the underlying molecular mechanism has not been analyzed yet. Thus, we named the autoregulatory mechanism controlling the capacity of the Golgi apparatus the "Golgi stress response" by analogy to the ER stress response, and started to analyze it. We are now trying to identify target genes whose transcription is induced by the Golgi stress response, and to clarify the mechanism of transcriptional regulation.

Conclusion

Here, I proposed to redefine the ER stress response as an autoregulatory system controlling the amount of the ER, and put forward the novel question "How is the amount of each organelle regulated in cells?" Although the question is so basic that researchers have taken it for granted or been indifferent to it, its analysis would be indispensable for understanding eukaryotic cells as molecular machinery. Moreover, analysis of the question may result in the revelation of interesting biological paradigms, such as the cytoplasmic mRNA splicing that was discovered during the study of the ER stress response (Ruegsegger, et al., 2001; Yoshida, et al., 2001a).

Acknowledgements

I thank Dr. Elizabeth Nakajima for critical reading of the manuscript, and Ms. Kaoru Miyagawa for secretarial assistance. This work was supported by the PRESTO-SORST program of the Japan Science and Technology Agency, and grants from the Ministry of Education, Culture, Sports, Science, and Technology (MEXT) of Japan (No. 18050013, No. 19370086 and No. 20052014). It was also financially supported in part by the Global Center of Excellence Program "Formation of a Strategic Base for Biodiversity and Evolutionary Research: from Genome to Ecosystem" of MEXT.

REFERENCES

Aldridge, J. E., Horibe, T. & Hoogenraad, N. J. (2007). Discovery of genes activated by the mitochondrial unfolded protein response (mtUPR) and cognate promoter elements. *PLoS ONE 2*, e874.

Benedetti, C., Haynes, C. M., Yang, Y., Harding, H. P. & Ron, D. (2006). Ubiquitin-like protein 5 positively regulates chaperone gene expression in the mitochondrial unfolded protein response. *Genetics 174*, 229-239.

Berger, E. G. & Roth, J. (1997). *The Golgi Apparatus* (Basel, Switzerland: Birkheuser Verlag).

Brodsky, J. L. (2007). The protective and destructive roles played by molecular chaperones during ERAD (endoplasmic-reticulum-associated degradation). *Biochem J 404*, 353-363.

Calfon, M., Zeng, H., Urano, F., Till, J. H., Hubbard, S. R., Harding, H. P., Clark, S. G. & Ron, D. (2002). IRE1 couples endoplasmic reticulum load to secretory capacity by processing the XBP-1 mRNA. *Nature 415*, 92-96.

Clermont, Y., Xia, L., Rambourg, A., Turner, J. D. & Hermo, L. (1993). Structure of the Golgi apparatus in stimulated and nonstimulated acinar cells of mammary glands of the rat. *Anat Rec 237*, 308-317.

Fujiki, Y. (2000). Peroxisome biogenesis and peroxisome biogenesis disorders. *FEBS Lett 476*, 42-46.

Fujiki, Y., Miyata, N., Matsumoto, N. & Tamura, S. (2008). Dynamic and functional assembly of the AAA peroxins, Pex1p and Pex6p, and their membrane receptor Pex26p involved in shuttling of the PTS1 receptor Pex5p in peroxisome biogenesis. *Biochem Soc Trans 36*, 109-113.

Gething, M.-J. (1997). *Guidebook to Molecular Chaperones and Protein-Folding Catalysts* (Oxford: Oxford University Press).

Green, S. & Wahli, W. (1994). Peroxisome proliferator-activated receptors: finding the orphan a home. *Mol Cell Endocrinol 100*, 149-153.

Gunn, K. E., Gifford, N. M., Mori, K. & Brewer, J. W. (2004). A role for the unfolded protein response in optimizing antibody secretion. *Mol Immunol 41*, 919-927.

Harding, H. P., Novoa, I. I., Zhang, Y., Zeng, H., Wek, R., Schapira, M. & Ron, D. (2000a). Regulated translation initiation controls stress-induced gene expression in mammalian cells. *Mol Cell 6*, 1099-1108.

Harding, H. P., Zhang, Y., Bertolotti, A., Zeng, H. & Ron, D. (2000b). Perk is essential for translational regulation and cell survival during the unfolded protein response. *Mol Cell 5*, 897-904.

Harding, H. P., Zhang, Y. & Ron, D. (1999). Protein translation and folding are coupled by an endoplasmic-reticulum-resident kinase. *Nature 397*, 271-274.

Harding, H. P., Zhang, Y., Zeng, H., Novoa, I., Lu, P. D., Calfon, M., Sadri, N., Yun, C., Popko, B., Paules, R., Stojdl, D. F., Bell, J. C., Hettmann, T., Leiden, J. M. & Ron, D. (2003). An integrated stress response regulates amino acid metabolism and resistance to oxidative stress. *Mol Cell 11*, 619-633.

Haynes, C. M., Petrova, K., Benedetti, C., Yang, Y. & Ron, D. (2007). ClpP mediates activation of a mitochondrial unfolded protein response in C. elegans. *Dev Cell 13*, 467-480.

Haze, K., Okada, T., Yoshida, H., Yanagi, H., Yura, T., Negishi, M. & Mori, K. (2001). Identification of the G13 (cAMP-response-element-binding protein-related protein) gene product related to activating transcription factor 6 as a transcriptional activator of the mammalian unfolded protein response. *Biochem J 355*, 19-28.

Haze, K., Yoshida, H., Yanagi, H., Yura, T. & Mori, K. (1999). Mammalian transcription factor ATF6 is synthesized as a transmembrane protein and activated by proteolysis in response to endoplasmic reticulum stress. *Mol Biol Cell 10*, 3787-3799.

Helenius, A. & Aebi, M. (2004). Roles of N-linked glycans in the endoplasmic reticulum. *Annu Rev Biochem 73*, 1019-1049.

Horibe, T. & Hoogenraad, N. J. (2007). The chop gene contains an element for the positive regulation of the mitochondrial unfolded protein response. *PLoS ONE 2*, e835.

Iwakoshi, N. N., Lee, A. H., Vallabhajosyula, P., Otipoby, K. L., Rajewsky, K. & Glimcher, L. H. (2003). Plasma cell differentiation and the unfolded protein response intersect at the transcription factor XBP-1. *Nat Immunol 4*, 321-329.

Iwawaki, T., Hosoda, A., Okuda, T., Kamigori, Y., Nomura-Furuwatari, C., Kimata, Y., Tsuru, A. & Kohno, K. (2001). Translational control by the ER transmembrane kinase/ribonuclease IRE1 under ER stress. *Nat Cell Biol 3*, 158-164.

Katayama, T., Imaizumi, K., Manabe, T., Hitomi, J., Kudo, T. & Tohyama, M. (2004). Induction of neuronal death by ER stress in Alzheimer's disease. *J Chem Neuroanat 28*, 67-78.

Lee, A. H., Chu, G. C., Iwakoshi, N. N. & Glimcher, L. H. (2005). XBP-1 is required for biogenesis of cellular secretory machinery of exocrine glands. *Embo J 24*, 4368-4380.

Lee, A. H., Iwakoshi, N. N. & Glimcher, L. H. (2003). XBP-1 regulates a subset of endoplasmic reticulum resident chaperone genes in the unfolded protein response. *Mol Cell Biol 23*, 7448-7459.

Lock, E. A., Mitchell, A. M. & Elcombe, C. R. (1989). Biochemical mechanisms of induction of hepatic peroxisome proliferation. *Annu Rev Pharmacol Toxicol 29*, 145-163.

Malhotra, J. D. & Kaufman, R. J. (2007). The endoplasmic reticulum and the unfolded protein response. *Semin Cell Dev Biol 18*, 716-731.

Martinus, R. D., Garth, G. P., Webster, T. L., Cartwright, P., Naylor, D. J., Hoj, P. B. & Hoogenraad, N. J. (1996). Selective induction of mitochondrial chaperones in response to loss of the mitochondrial genome. *Eur J Biochem 240*, 98-103.

Mori, K. (2003). Frame switch splicing and regulated intramembrane proteolysis: key words to understand the unfolded protein response. *Traffic 4*, 519-528.

Oku, M., Warnecke, D., Noda, T., Muller, F., Heinz, E., Mukaiyama, H., Kato, N. & Sakai, Y. (2003). Peroxisome degradation requires catalytically active sterol glucosyltransferase with a GRAM domain. *Embo J 22*, 3231-3241.

Oyadomari, S., Araki, E. & Mori, M. (2002). Endoplasmic reticulum stress-mediated apoptosis in pancreatic beta-cells. *Apoptosis 7*, 335-345.

Platta, H. W. & Erdmann, R. (2007). Peroxisomal dynamics. *Trends Cell Biol 17*, 474-484.

Rambourg, A., Clermont, Y., Chretien, M. & Olivier, L. (1993). Modulation of the Golgi apparatus in stimulated and nonstimulated prolactin cells of female rats. *Anat Rec 235*, 353-362.

Reimold, A. M., Iwakoshi, N. N., Manis, J., Vallabhajosyula, P., Szomolanyi-Tsuda, E., Gravallese, E. M., Friend, D., Grusby, M. J., Alt, F. & Glimcher, L. H. (2001). Plasma cell differentiation requires the transcription factor XBP-1. *Nature 412*, 300-307.

Reimold, A. M., Ponath, P. D., Li, Y. S., Hardy, R. R., David, C. S., Strominger, J. L. & Glimcher, L. H. (1996). Transcription factor B cell lineage-specific activator protein regulates the gene for human X-box binding protein 1. *J. Exp. Med. 183*, 393-401.

Ron, D. & Walter, P. (2007). Signal integration in the endoplasmic reticulum unfolded protein response. *Nat Rev Mol Cell Biol 8*, 519-529.

Rottensteiner, H., Wabnegger, L., Erdmann, R., Hamilton, B., Ruis, H., Hartig, A. & Gurvitz, A. (2003). Saccharomyces cerevisiae PIP2 mediating oleic acid induction and peroxisome proliferation is regulated by Adr1p and Pip2p-Oaf1p. *J Biol Chem 278*, 27605-27611.

Ruddock, L. W. & Molinari, M. (2006). N-glycan processing in ER quality control. *J Cell Sci 119*, 4373-4380.

Ruegsegger, U., Leber, J. H. & Walter, P. (2001). Block of HAC1 mRNA translation by long-range base pairing is released by cytoplasmic splicing upon induction of the unfolded protein response. *Cell 107*, 103-114.

Ryan, M. T. & Hoogenraad, N. J. (2007). Mitochondrial-nuclear communications. *Annu Rev Biochem 76*, 701-722.

Sakai, Y., Oku, M., van der Klei, I. J. & Kiel, J. A. (2006). Pexophagy: autophagic degradation of peroxisomes. *Biochim Biophys Acta 1763*, 1767-1775.

Sekine, Y., Takeda, K. & Ichijo, H. (2006). The ASK1-MAP kinase signaling in ER stress and neurodegenerative diseases. *Curr Mol Med 6*, 87-97.

Shaffer, A. L., Shapiro-Shelef, M., Iwakoshi, N. N., Lee, A. H., Qian, S. B., Zhao, H., Yu, X., Yang, L., Tan, B. K., Rosenwald, A., Hurt, E. M., Petroulakis, E., Sonenberg, N., Yewdell, J. W., Calame, K., Glimcher, L. H. & Staudt, L. M. (2004). XBP1, downstream of Blimp-1, expands the secretory apparatus and other organelles, and increases protein synthesis in plasma cell differentiation. *Immunity 21*, 81-93.

Sriburi, R., Jackowski, S., Mori, K. & Brewer, J. W. (2004). XBP1: a link between the unfolded protein response, lipid biosynthesis, and biogenesis of the endoplasmic reticulum. *J Cell Biol 167*, 35-41.

Takahashi, R., Imai, Y., Hattori, N. & Mizuno, Y. (2003). Parkin and endoplasmic reticulum stress. *Ann N Y Acad Sci 991*, 101-106.

Tan, Y., Dourdin, N., Wu, C., De Veyra, T., Elce, J. S. & Greer, P. A. (2006). Ubiquitous calpains promote caspase-12 and Jnk activation during ER stress-induced apoptosis. *J Biol Chem 281*, 16016-16024.

Tirasophon, W., Welihinda, A. A. & Kaufman, R. J. (1998). A stress response pathway from the endoplasmic reticulum to the nucleus requires a novel bifunctional protein kinase/endoribonuclease (Ire1p) in mammalian cells. *Genes Dev 12*, 1812-1824.

Urano, F., Wang, X., Bertolotti, A., Zhang, Y., Chung, P., Harding, H. P. & Ron, D. (2000). Coupling of stress in the ER to activation of JNK protein kinases by transmembrane protein kinase IRE1. *Science 287*, 664-666.

Yamamoto, K., Sato, T., Matsui, T., Sato, M., Okada, T., Yoshida, H., Harada, A. & Mori, K. (2007). Transcriptional induction of mammalian ER quality control proteins is mediated by single or combined action of ATF6alpha and XBP1. *Dev Cell 13*, 365-376.

Yamashita, S., Oku, M. & Sakai, Y. (2007). Functions of PI4P and sterol glucoside are necessary for the synthesis of a nascent membrane structure during pexophagy. *Autophagy 3*, 35-37.

Ye, J., Rawson, R. B., Komuro, R., Chen, X., Dave, U. P., Prywes, R., Brown, M. S. & Goldstein, J. L. (2000). ER stress induces cleavage of membrane-bound ATF6 by the same proteases that process SREBPs. *Mol Cell 6*, 1355-1364.

Yoneda, T., Benedetti, C., Urano, F., Clark, S. G., Harding, H. P. & Ron, D. (2004). Compartment-specific perturbation of protein handling activates genes encoding mitochondrial chaperones. *J Cell Sci 117*, 4055-4066.

Yoshida, H. (2007). ER stress and diseases. *Febs J 274*, 630-658.

Yoshida, H., Haze, K., Yanagi, H., Yura, T. & Mori, K. (1998). Identification of the cis-acting endoplasmic reticulum stress response element responsible for transcriptional induction of mammalian glucose-regulated proteins. Involvement of basic leucine zipper transcription factors. *J Biol Chem 273*, 33741-33749.

Yoshida, H., Matsui, T., Hosokawa, N., Kaufman, R. J., Nagata, K. & Mori, K. (2003). A time-dependent phase shift in the mammalian unfolded protein response. *Dev Cell 4*, 265-271.

Yoshida, H., Matsui, T., Yamamoto, A., Okada, T. & Mori, K. (2001a). XBP1 mRNA is induced by ATF6 and spliced by IRE1 in response to ER stress to produce a highly active transcription factor. *Cell 107*, 881-891.

Yoshida, H., Okada, T., Haze, K., Yanagi, H., Yura, T., Negishi, M. & Mori, K. (2000). ATF6 activated by proteolysis binds in the presence of NF-Y (CBF) directly to the cis-acting element responsible for the mammalian unfolded protein response. *Mol Cell Biol 20*, 6755-6767.

Yoshida, H., Okada, T., Haze, K., Yanagi, H., Yura, T., Negishi, M. & Mori, K. (2001b). Endoplasmic reticulum stress-induced formation of transcription factor complex ERSF including NF-Y (CBF) and activating transcription factors 6alpha and 6beta that activates the mammalian unfolded protein response. *Mol Cell Biol 21*, 1239-1248.

Zhao, Q., Wang, J., Levichkin, I. V., Stasinopoulos, S., Ryan, M. T. & Hoogenraad, N. J. (2002). A mitochondrial specific stress response in mammalian cells. *Embo J 21*, 4411-4419.

In: Protein Misfolding
Editors: C. B. O'Doherty, A.C. Byrne

ISBN: 978-1-60456-881-3

Chapter VIII

Targeting Stress Pathways to Treat Neurological Disorders Linked to Protein Misfolding

Soledad Matus, Peter Thielen, Melissa Nassif, Mauricio Torres and Claudio Hetz*

Program of Cellular and Molecular Biology, Institute of Biomedical Sciences
Faculty of Medicine and The FONDAP Center for Molecular Studies of the Cell,
University of Chile, Santiago, Chile

Abstract

Neurodegenerative disorders such as Amyotrophic lateral sclerosis (ALS), Alzheimer's disease, Parkinson's disease, Prion-related disorders (PrD) and Huntington's disease (HD) share a common neuropathology, primarily featuring the presence of abnormal protein inclusions containing specific misfolded proteins. Recent evidence indicates that alteration in organelle function is a common pathological feature of protein misfolding disorders. The endoplasmic reticulum (ER) is an essential compartment for protein folding, maturation, and secretion. Signs of ER stress have been extensively described in most experimental models of neurological disorders and more recently in the brains of human patients. ER stress is caused by functional disturbances which result in accumulation of unfolded/misfolded proteins at the ER lumen. To cope with ER stress, cells activate an integrated signaling response termed the Unfolded Protein Response (UPR) which aims to reestablish homeostasis through transcriptional upregulation of genes involved in folding, protein quality control and protein degradation pathways including lysosome-mediated autophagy. Small molecules with chaperone-like activity have been shown to alleviate ER stress and decrease protein misfolding in disease

* Correspondence concerning this article should be addressed to: Claudio Hetz, Independencia 1027, Faculty of Medicine, Institute of Biomedical Sciences, University of Chile, Santiago, Chile. P.O.Box 70086. Phone: 56-29786506 Email: chetz@med.uchile.cl. Website: http://www.annabolteus.com/HetzLab/Index.html.

settings. In this chapter we give an overview of the role of ER stress in pathological conditions such as protein misfolding disorders and brain ischemia, and discuss possible pharmacological strategies to target the UPR with therapeutic benefits.

INTRODUCTION

Most neurodegenerative disorders share a common neuropathology, primarily featuring the presence of abnormal protein inclusions containing specific misfolded proteins. These diseases are now classified as Protein Misfolding Disorders and include neuropathological conditions such as Amyotrophic lateral sclerosis (ALS), Alzheimer's disease, Parkinson's disease, Prion-related disorders (PrD), Huntington's disease (HD), and many others [1-4]. Disease-related misfolded proteins alter essential cellular functions that lead to neurological impairment and, in many cases, cellular death. General alterations include synapse abnormalities, alteration in axonal transport, redox changes, proteasome dysfunction, and cytoskeleton disorganization. Increasing evidence indicates that organelle stress is an important pathological target in neurodegeneration. Special attention has been given in the last years to one subcellular organelle, the endoplasmic reticulum (ER), in brain disorders (Table 1). The ER is a specialized subcellular compartment essential for the folding of proteins destined for the secretory pathway. In most protein misfolding disorders, upregulation of markers of ER stress is commonly observed related to the activation of an adaptive reaction known as the Unfolded Protein Response (UPR). The UPR aims to reestablish homeostasis by restoring the cells capacity to produce properly folded proteins. In doing so, activation of the UPR affects expression of proteins involved in nearly every aspect of the secretory pathway, including folding, ER-associated degradation, protein entry to the ER, as well as other adaptive responses such as autophagy. Under chronic or irreversible ER damage the UPR will ultimately initiate apoptosis to eliminate damaged cells. Although ER stress is observed in neurological disorders, the actual contribution of the pathway to the disease process remains speculative and most available data is either correlative or based on *in vitro* evidence. This chapter focuses on recent findings implicating the activation of the UPR in pathological conditions affecting the nervous system and the possible use of pharmacological strategies to target this pathway with small molecules with therapeutic benefits.

I. UPR SIGNALING: COMPONENTS AND CONSEQUENCES

The primary function of the ER is to facilitate protein folding and secretion. A complex system of protein chaperones is employed to catalyze the efficient folding of proteins and prevent abnormal aggregation or misfolding of proteins. This organelle is responsible for regulating and executing many posttranslational modifications, ensuring proper protein function and facilitating the formation of functional protein complexes. The ER also serves as the major calcium store, and biosynthesis of steroids, cholesterol, and other lipids occurs within, playing a crucial role in organelle biogenesis and signaling. A number of stress conditions can interfere with its function and therefore lead to abnormal protein folding in the ER lumen. Accumulation of unfolded and/or misfolded proteins causes an imbalance between the synthesis of new proteins and the ER's ability to process newly synthesized proteins, resulting in failure of the ER to cope with excessive protein load, a condition termed 'ER stress' [5]. As a result, cells activate an integrated intracellular signaling cascade, the UPR, to avert ER stress. Activation of the UPR results in an attenuation of the rate of protein synthesis, upregulation of genes encoding chaperones, foldases, proteins related to maturation, and proteins involved in the retro-translocation and degradation of ER-generated misfolded proteins by the proteasome. These responses are initiated to minimize accumulation and aggregation of misfolded proteins by increasing the functional capacity of the ER to facilitate folding and to degrade abnormal proteins [6,7].

Table 1. ER stress and neurodegeneration. The table summarizes information obtained from animal and cellular models of neurodegeneration. Experimental validation for activation of the UPR in human patients affected with neurological diseases is also indicated from analysis of post-mortem samples

Protein Conformational Disorder	ER stress UPR activation	UPR markers	Human patients
Alzheimer	In vitro and in vivo	✓	✓
Parkinson	In vitro and in vivo	✓	✓
Amyotrophic Lateral Sclerosis	In vitro and in vivo	✓	✓
Prion disorders (TSEs)	In vitro and in vivo	✓	✓
Retroviral Spongiform degeneration	In vitro and in vivo	✓	
Huntington	In vitro	✓	
Kennedy disease	In vitro	✓	
Spinicerebellar Ataxias	In vitro	✓	
Vanishing White Matter Disease	ND	✓	✓
Brain Ischemia/Brain trauma	In vitro and in vivo	✓	
Brain Ischemia/ stroke	In vitro and in vivo	✓	

There are three main ER resident transmembrane proteins that act as stress sensors to initiate different UPR signaling cascades: double-stranded RNA-activated protein kinase (PKR)-like endoplasmic reticulum kinase (PERK), activating transcription factor 6 (ATF6) alpha and beta, and inositol requiring kinase 1 (IRE1) alpha and beta (Figure 1). Each of these stress sensors transduces information of the protein folding status at the ER to the nucleus by controlling expression of specific transcription factors. Thus, PERK, ATF6, and IRE1a operate in concert to ensure adaptation to protein folding stress (reviewed in [8]).

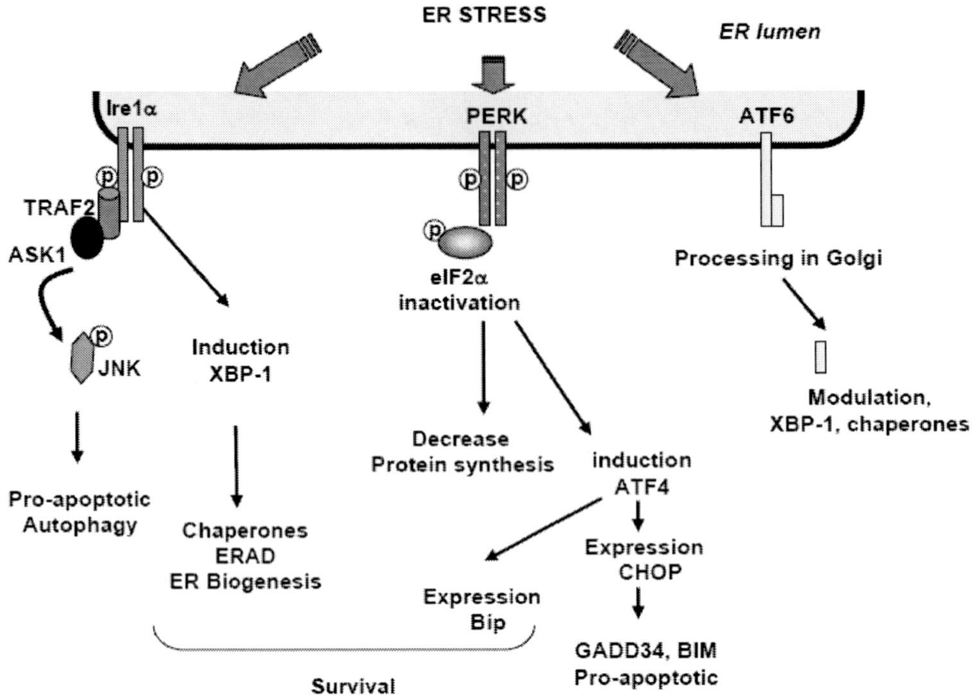

Figure 1. The unfolded protein response. Accumulation of misfolded protein inside the endoplasmic reticulum (ER) lumen triggers a stress response known as the unfolded protein response (UPR). In cells undergoing ER stress, IRE1 auto-phosphorylates, leading to the activation of its endoribonuclease activity. This activity mediates the processing of the mRNA encoding *XBP-1*, which is a transcription factor that upregulates many essential UPR genes involved in folding and protein quality control. Alternatively, activation of PERK increases the translation of *ATF4*, a transcription factor that induces the expression of genes such as CHOP that function in amino acid metabolism, antioxidant response, and apoptosis. A third UPR pathway is initiated by ATF6, a type II ER transmembrane protein encoding a bZIP transcriptional factor on its cytosolic domain and localized in the ER in unstressed cells. Upon ER stress induction, ATF6 is processed, increasing the expression of some ER chaperones.

Activation of PERK is mediated by its oligomerization and autophosphorylation, leading to the phosphorylation and inhibition of eukaryotic translation initiation factor 2α (eIF2α) [9]. Alternatively, eIF2a phosphorylation augments the specific translation of ATF4, a UPR transcription factor essential for the upregulation of many essential UPR genes such as CHOP and Grp78/BiP. Activated ATF6 translocates from the ER to the Golgi apparatus where it is proteolytically processed, releasing its cytosolic domain which is then translocated to the nucleus to function as an active transcription factor. This cleaved form of ATF6 upregulates several ER chaperones (Figure 1) [10-12].

The third adaptive response is regulated by IRE1α and its downstream target X-Box-binding protein 1 (XBP-1). IRE1α is a Serine/Threonine protein kinase and endoribonuclease that, upon activation, initiates the unconventional splicing of the mRNA encoding the transcription factor XBP-1 [13-15] (Figure 1). A 26 nucleotide intron of XBP-1 mRNA is removed by activated IRE1α, resulting in an alternative splice form which shifts the mRNA reading frame. This mRNA maturation event promotes the expression of a more stable and potent transcriptional activator, XBP-1s, that controls the upregulation of a subset of UPR-related genes (Figure 1A). Through this process IRE1α transduces survival signals to the nucleus to increase the folding capacity of the ER, therefore alleviating stress. In addition, activated IRE1a binds the adaptor protein TRAF2, leading to the activation of several signaling pathways including ERK, JNK and NF-kB [6]. The signaling activity of IRE1α is modulated by the formation of a protein complex termed the *UPRosome* (reviewed in [8,16]), which is controlled by the binding of different accessory proteins including some members of the BCL-2 protein family and other components [16,17].

II. Protein Misfolding and Neurodegeneration

As previously mentioned, a common feature of many neurodegenerative diseases is the accumulation and deposition of misfolded proteins that affects neuronal function and viability [2,18-20]. Experimental evidence using mouse models of neurodegeration indicate that several mutations associated with hereditary forms of disease activate various components of the UPR (Table 1), implying impaired ER function (Figure 2). More importantly, upregulation of ER stress markers has been observed in post-mortem brain tissues and cell culture models of many protein conformational disorders, including Parkinson's [21-23], ALS [24,25], Alzheimer's [26] and Creutzfeldt-Jacob's disease [27,28]. The expected functional significance of ER stress to the disease process is proposed to be both a protective component during early UPR responses and more deleterious during prolonged response due to disturbance of ER homeostasis (reviewed in [29,30]). In the following sections we discuss specific evidence linking ER stress to neurodegeneration in three diseases where therapeutic strategies have been employed to target ER stress responses: ALS, HD and PrD. In addition, experimental evidence linking ER stress with brain ischemia and stroke conditions is presented.

Huntington's Disease

HD is an autosomal dominant neurodegenerative disorder characterized by progressive cognitive impairment, neuropsychiatric symptoms, and chorea. HD is caused by a mutation encoding an abnormal expansion of CAG-encoded polyglutamine repeats within the Huntingtin protein. Onset depends on the severity of the mutation, usually resulting in symptoms beginning in the fourth or fifth decade of life. Patients typically decline over a 15-20 year period until complications lead to death, as there is no effective treatment for the selective neuronal loss affecting the frontal lobe, striatum, and basal ganglia [31]. The general population exhibits an average of 18 poly(Q) repeats, yet expansions exceeding 35 repeats will almost always result in disease. Increasing numbers of repeats cause the age of onset to decrease, and individuals with more than 60 repeats will develop disease before the age of twenty [32].

The primary histopathological feature observed in HD is the co-localization of Huntingtin positive-aggregates with ubiquitin [33]. In many cases, the formation of intracellular Huntingtin inclusions precede cell death [34-36], and neurotoxicity is also plausible when considering the array of other proteins found within such aggregates [37,38]. Changes in subcellular localization of Huntingtin-interacting proteins may also contribute to the pathology [39].

Although the mechanism of expanded poly(Q) pathogenesis is still highly controversial, it is well established that neurotoxicity is provoked by a dominant gain of function. Neuronal apoptosis in HD is proposed to be mediated by several different mechanisms, including mitochondrial dysfunction, altered axonal transport, and proteasome dysfunction, to name a few (reviewed in [40]). Recent data also suggests that pro-apoptotic pathways originating from the ER may contribute to HD neurodegeneration. Mutant Huntingtin aggregates impair the function of the ubiquitin-proteasome system [40], resulting in ER stress through general accumulation of abnormally folded proteins, poisoning the cell with abnormal folding intermediates or immature proteins [41].

Recent attempts to understand the function of wild type Huntingtin demonstrated that inhibition of its expression significantly alters ER morphogenesis and network patterning [42]. In addition, mutant Huntingtin perturbs ER calcium homeostasis [43], and the experimental targeting of poly(Q) peptides to the ER decreases aggregation [44]. Thus, increasing evidence suggests that mutant Huntingtin may exert its neurotoxic effect by directly altering ER function. However, the actual involvement of ER stress-related pathways in the disease remains speculative, as no *in vivo* experiments have validated these findings.

The accumulation of mutant poly(Q) inclusions triggers activation of UPR stress sensors IRE1α and PERK *in vitro*, resulting in upregulation of downstream targets [45]. Many groups have shown that ER stress partially mediates toxicity caused by mutant HD associated with activation of JNK, ASK1 and caspase-12 processing, an ER located caspase [46]. ASK1 deficient neurons have been shown to protect from poly$(Q)_{79}$ toxicity [136,137,141], [47]. One of primary roles of the UPR is to induce expression of a diverse set of chaperones, and many reports have shown that mutant Huntingtin aggregation can be diminished by the expression of chaperones such as Hsp70 and Hsp40 [36,48].

Evidence is beginning to emerge indicating that clearance pathways responsible for elimination of poly(Q) aggregates are closely regulated by the ER through UPR stress sensors such as PERK/eIF2α, which control expression of the autophagy-related gene ATG12 [45]. Thus, the UPR and ER stress pathways may affect the accumulation of mutant Huntington in the cell. Autophagy is a "large-scale" cellular degradation process for proteins and damaged organelles (Figure 2) [49]. Autophagy is divided in three subtypes, including macroautophagy, microautophagy, and chaperone-mediated autophagy. This review will focus on macroautophagy, hereafter referred to as autophagy. During autophagy, double membrane vesicles termed autophagosomes are formed, sequestering cytosolic proteins and/or organelles as cargoes (Figure 2). Autophagosomes then fuse with lysosomes, and their intracellular components are degraded to ultimately result in macromolecule recycling.

The therapeutic effects of targeting the UPR were recently demonstrated in models of diabetes where ER stress mediates insulin resistance [50]. The authors employed chemical chaperones, including 4-phenyl butyric acid (4-PBA) and trimethylamine N-oxide dihydrate (TMAO), which are a group of low molecular weight compounds known to stabilize protein conformation and improve ER folding capacity (reviewed in [51]). Likewise, endogenous bile acid derivates, such as TUDCA, also modulate ER function [52]. Remarkably, a positive effect of TUDCA administration was demonstrated on different models of Huntington's disease [53], reducing striatal-neuron apoptosis and the size of intranuclear Huntingtin inclusions. Locomotor and sensorimotor deficits were significantly improved in HD animal models upon treatment [53]. The chemical chaperone 4-PBA also significantly extended survival, improved motor performance, and delayed the neuropathological features in the R6/2 transgenic mouse model of HD [54]. The impact these drugs have on ER stress in HD transgenic mouse models remains to be determined.

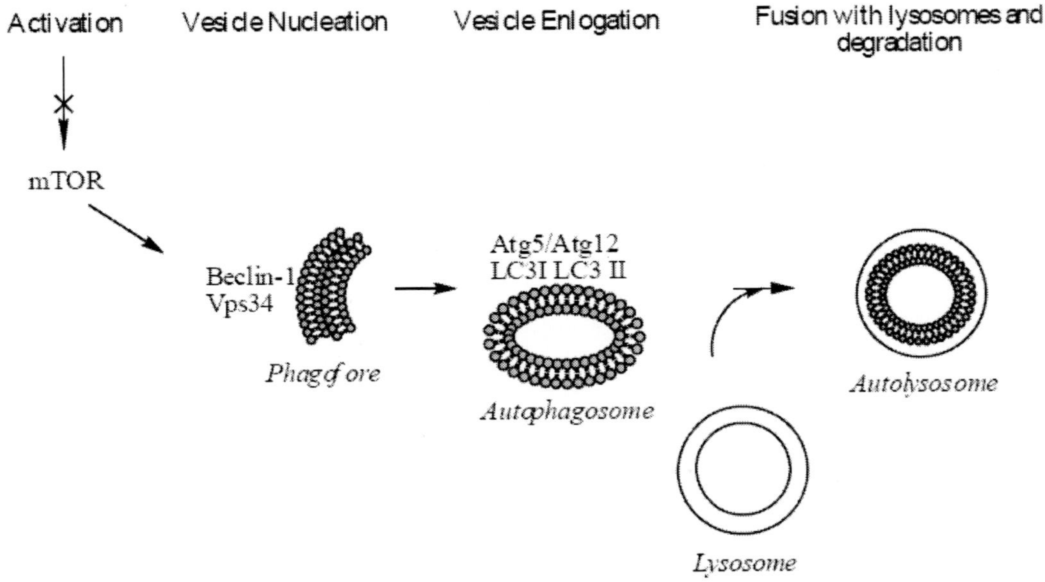

Figure 2. Autophagy. Autophagy can be induced for several stress stimuli including ER stress, nutrient starvation and accumulation of protein aggregates. Different proteins regulate the initiation of autophagy, and the formation and activation of kinase complexes with Atg proteins triggers nucleation and elongation leading to formation of the double-membrane autophagosome. Beclin-1 also mediates the initiation of autophagy and is negatively regulated through an interaction with BCL-2 at the ER membrane. Inhibition of the mammalian target of rapamycin (mTOR) activates autophagy. Finally, the autophagosome matures by fusing with endosomes and lysosomes forming the autophagolysosome where the internal material is degraded, resulting in macromolecule recycling. Phosphatidylethanolamine (PE)-conjugated LC3-II is the most used marker that specifically localizes to autophagosomes.

The ability of chemical chaperones to alleviate ER stress *in vitro* demonstrates the feasibility of targeting organelle function for therapeutic gain. Accumulating evidence indicates a constant homeostatic cross-talk between the UPR and autophagy [55]. In this line, activation of autophagy with rapamycin, which targets the mTor pathway, has the ability to increase the survival of HD animal models and recover motor performance. Autophagy occurs at basal levels in most tissues [56-58], yet activation preferentially occurs mainly under two different stress conditions: during starvation it acts to promote survival, and in times when misfolded proteins or damaged organelles are present it acts to rid the cell of altered intracellular components. The therapeutic effects of rapamycin in HD models were associated with a decrease in Huntingtin aggregate size, indicating that pharmacological strategies aiming to increase autophagy-mediated clearance of protein aggregates may be clinically effective [59]. A relatively new drug, trehalose, was recently shown to act both as a chemical chaperone and an autophagy activator, decreasing the aggregation of mutant Huntingtin and alpha–synuclein *in vitro* [60]. The same group recently performed a small-molecule screen in yeast which identified novel small-molecule modulators of mammalian autophagy based on their ability to modulate the cytostatic effects of rapamycin [61]. Three enhancers induced autophagy independently of rapamycin in mammalian cells, facilitating the clearance of mutant Huntingtin and Parkinson's related a-synuclein^{A53T}. These autophagy enhancers, which seem to act either independently or downstream of rapamycin, attenuated mutant Huntingtin-fragment toxicity in Huntington's disease *in vitro* models as well as in *D. melanogaster* [62], reinforcing the therapeutic potential for clearance of mutant protein aggregates identified in neurodegenerative conditions.

Amyotrophic Lateral Sclerosis

ALS, also known as Lou Gehrig's disease, is a progressive and deadly adult-onset motor neuron disease characterized by muscle weakness, spasticity, atrophy, paralysis and premature death. The pathological hallmark of ALS is the selective degeneration of motoneurons in the spinal ventral horn, most brainstem nuclei and the cerebral cortex. Average age of onset occurs around 50 years of age, with an incidence estimated at 1-2 per 100,000 individuals, males being more frequently affected than females. The majority of ALS patients lack a defined hereditary genetic component and are termed sporadic (sALS), while approximately 10% of cases are familial (fALS). Of these, over 100 mutations affect the gene encoding superoxide dismutase-1 (SOD1), triggering misfolding and abnormal aggregation in which leads to neurotoxicity. Overexpression of human fALS-linked SOD1 mutations in transgenic mice recapitulates essential features of the human pathology, provoking age-dependent protein aggregation, paralysis and motor neuron degeneration (see general reviews in [63,64]). Since sporadic and familial ALS affect the same neurons with similar pathology, therapeutics effective in mutant SOD1 models may translate to sporadic ALS.

The primary mechanism by which mutations in SOD1 contribute to progressive motoneuron loss in fALS remains unknown, but it has been proposed that neuronal loss is mediated by mitochondrial dysfunction, altered axonal transport, and non-neuronal components (reviewed in [63,64]). Recent data also suggests that stress responses originating from the endoplasmic reticulum (ER) may contribute to ALS [65]. Activation of the UPR was recently described in human post-mortem samples from fALS and sALS patients [24,25]. Similarly, ER stress responses are observed in transgenic rodents expressing different fALS-related SOD1 mutations [66-73] and biochemical and histological studies revealed that a fraction of mutant SOD1 accumulates inside the ER and Golgi *in vivo*, where it forms insoluble high molecular weight species [68,70,71]. A direct interaction between mutant SOD1 and the ER chaperones Grp78/BiP and PDI was observed in microsomal fractions of spinal cord extracts [68,70]. Significantly, a proteomic analysis of spinal cord tissue from SOD1^{G93A} transgenic mice revealed that two UPR targets, PDI and Grp58, are among the most induced proteins in this ALS model [70]. Moreover, involvement of pro-apoptotic BIM and PUMA, both essential for ER stress-induced apoptosis [74,75], in motoneuron loss and disease progression has recently been described in mouse models of fALS [73,76],. Although a strong correlation between ER stress, SOD1 aggregation and ALS progression has been described by many independent groups, the actual role of the UPR in the disease process has not been addressed directly and remains speculative.

Several pieces of evidence suggest that chronic or irreversible ER stress may play an important role in motoneuron loss in ALS. Many groups have described upregulation of ER stress related pro-apoptotic factors in the spinal cord of late stage disease animals, including GADD153/CHOP and pro-caspase-12 processing [66,69,70,72,73]. In addition, ER stress inducible BCL-2 pro-apoptotic genes such as BIM and PUMA are upregulated in symptomatic mutant SOD1 transgenic mice [73,76]. More importantly, genetic deletion of BIM or PUMA delays ALS disease onset, possibly due to a significant decrease of apoptosis in motoneurons. This data is in agreement with the fact that BAX deficient or BCL-2 transgenic mice exhibit an increased life span due to augmented motoneuron survival [77,78].

In support of the role of ER stress on ALS neuropathological, salubrinal, a small molecule described as an ER stress inhibitor [79], was shown to suppress aggregation of mutant SOD1 as well as SOD1-induced apoptotic cell death in Neuro2A cells [80]. Salubrinal selectively engages the translational control branch of the UPR by inducing eIF2a phosphorylation without affecting the transcription-dependent component of the UPR [79]. However, it remains to be shown that this protective effect is due to the delay of aggregate formation or inhibition of downstream ER stress responses caused by mutant SOD1 oligomers.

Neuroprotective therapies that target specific molecular mechanisms in ALS mice, including expression of anti-apoptotic genes and growth factors through viral-mediated gene therapy, have potential to delay onset and slow progression of disease. In addition, targeting of mutant SOD1 with small interfering RNA and lentiviral vectors has therapeutic potential in animal models of ALS [81], and small molecules have also been shown to have therapeutic benefits [82-84]. 4-PBA was recently shown to decrease motoneuron apoptosis in SOD1 transgenic mice, improving motor performance and life span. However, the effects of 4-PBA as a chemical chaperone were not evaluated (ref.) The question remains as to how cells manage to accumulate mutant SOD1. Kabuta and colleagues reported that activation of autophagy reduces SOD1^{G93A} mediated toxicity and overall protein levels in a neuroblastoma cell line, proposing that the contribution of autophagy to mutant SOD1 degradation was comparable to that of proteasome pathway [85]. Increased levels of LC3 processing and inhibition of mTOR were observed in SOD1^{G93A} transgenic mice [86], possibly inferring that autophagy has a relevant role as a mechanism to eliminate SOD1 aggregates.

Based on these outcomes, therapeutic benefits of targeting autophagy are becoming more attractive for treatment of ALS. In a recent work the use of lithium, a recognized compound for treatment of depression and bipolar disorder, delayed the progression of the disease in human ALS patients [87]. Lithium treatment significantly delayed cell death of spinal cord neurons in SOD1^{G93A} transgenic mice, improving motor performance and extending life span. Interestingly, lithium produced a marked regression of SOD1 aggregates, decreased levels of ubiquitinated proteins, and reduced alpha-synuclein expression, a known component of the autophagy pathway, resulting in an increased number of autophagosomes. In an mTOR independent manner, lithium acts as a positive modulator of autophagy by blocking IP3 activity [88], possibly explaining its protective effect in ALS. Clinical trials are now underway to test the efficacy of treating ALS with lithium, and slower disease progression has been observed in comparison with control individuals. While these trials are still preliminary, it is encouraging that evidence gained from fALS mouse models is capable of translating to treatment of disease in sALS human patients. Large clinical trials are now required to establish the safety and efficacy of treating with lithium and other autophagy initiating therapeutics to reduce the progression of ALS.

Prion-Related Disorders

PrDs, also known as transmissible spongiform encephalopathies (TSEs), are fatal neurodegenerative disorders characterized by spongiform degeneration of the brain accompanied by accumulations of a misfolded and partially protease-resistant form of the prion protein (PrP^{Sc} or PrP^{RES}) [89]. Etiologically, TSEs can be classified as sporadic, infectious, or autosomal dominant inherited forms, affecting both humans and other mammals. In humans, Prion disorders include classic Creutzfeldt-Jakob disease (CJD), new variant CJD, Gerstmann-Sträussler-Scheinker syndrome, fatal familial insomnia and kuru. Prion diseases in other mammals include bovine spongiform encephalopathy, commonly known as mad cow disease, chronic wasting disease in elk and deer, and scrapie in sheep and goats [90]. This group of diseases generally presents a long incubation period followed by rapid progression to death shortly after the first clinical signs appear. Although clinical symptoms and progression vary among PrD, they commonly include progressive dementia and ataxia as a result of extensive neuronal death due to reactive astrocytosis and activated microglia during neuro-inflammation [91].

The most widely accepted "protein-only" hypothesis postulates that infectious prion pathogenicity results from a conformational change of normal cellular prion protein (PrP^{C}) from its primarily α-helical structure to an insoluble β sheet conformation (PrP^{Sc}), initiated by an direct interaction between the two PrP forms. No amino acid sequence or posttranslational modification differences have been detected between normal host PrP^{C} and its pathological form, PrP^{Sc} [92]. Like other membrane proteins, PrP^{C} undergoes post translational processing in the ER and Golgi. It then localizes to cholesterol-rich lipid rafts, where it is anchored to the outer surface of the plasma membrane with a glycosyl phosphatidyl inositol (GPI) anchor [93]. Despite the vast body of research, the physiological role of PrP^{C} is still under debate [94] (Figure ?).

Apoptosis has been described in the brains of patients affected by CJD and Fatal Familial Insomnia, as well as scrapie-infected hamsters, mice and sheep (reviewed in [95]). Several *in vitro* and *in vivo* models for PrD have been developed to understand the molecular basis of neuronal dysfunction in infectious and inherited forms of PrD [96-99]. However, the exact mechanism by which Prion misfolding affects neuronal function is not well understood. Several groups have shown an engagement of ER stress responses in PrD models, where XBP-1 splicing [100] and the activation of stress signaling pathways linked to ER stress, such as JNK and ERK, are observed in scrapie-infected mice [101,102]. In both human CJD patients and mouse models, the upregulation of UPR-responsive chaperones such as Grp78/BiP, Grp94, and Grp58 is observed in the brain [27,103]. In addition, a proteomic analysis of post-mortem CJD brain samples demonstrated that the disulfide isomerase Grp58 was highly expressed in the cerebellum of humans patients affected with sporadic CJD [28]. *In vitro* studies using Neuro2a neuroblastoma cells demonstrated that inhibition of Grp58 expression using RNA interference leads to a significant enhancement of PrP^{Sc} toxicity. Conversely, overexpression of Grp58 protects cells from PrP^{Sc} toxicity and decreases the rate of caspase-12 activation, suggesting that ER stress is an early protective cellular response to prion replication, acting as a neuroprotective mechanism against prion neurotoxicity [104]. Recent evidence indicates that the activation of UPR components such as IRE1α,, XBP-1 and others determines the rate of PrP^C aggregation *in vitro* under stress conditions [105,106] and in yeast models [107], suggesting that the UPR has an active role in preventing neurodegeneration. We recently described conditions in which ER stress facilitates *in vitro* conversion of PrP^C into PrP^{Sc} due to partial misfolding of PrP^C [105], which was reverted by activation of UPR components such as XBP-1, ATF4 and ATF6. In addition, scrapie infected neuroblastoma cells are more sensitive to ER stress mediated apoptosis [27]. Because PrP^{Sc} induces ER stress, it seems that a vicious cycle exists in which alterations of ER physiology cause PrP^C to be more susceptible to conversion into PrP^{Res}, therefore causing additional ER stress. This most likely explains the rapid progression of the disease.

Despite the fact that chronic ER stress has been extensively described in neurodegenerative conditions linked to protein misfolding and aggregation, the role of the UPR in the central nervous system had not been studied directly. To address this question, we recently described the generation of a brain specific XBP-1 conditional knockout strain ($XBP\text{-}1^{Nes-/-}$) and assessed the function of XBP-1 prion pathogenesis [108]. To our surprise, the activation of stress responses triggered by prion replication, such as Grp58 and PDI induction, caspase-12 processing, JNK and ERK phosphorylation, were not influenced by XBP-1 deficiency. Neither prion aggregation, neuronal loss nor animal survival were affected. These findings imply that this highly conserved arm of the UPR may not contribute to the occurrence or pathology of neurodegenerative conditions associated with Prion misfolding, despite predictions that such diseases are related to ER stress. Alternatively, one may speculate that the upregulation of ER chaperones is due to specific regulation of their promoters and not UPR activation. Since the UPR is not restricted to the IRE1α/XBP-1 pathway in mammals, an activation of other UPR pathways may possibly compensate for XBP-1 deficiency in our prion model. The contribution of ATF6 and PERK to prion pathogenesis remains to be determined using genetic manipulation *in vivo*.

Similarly, Steele *et al* recently described that caspase-12 knockout mice are normally susceptible to prion infection and pathogenesis despite a clear activation of the UPR in their model [109]. Pro-apoptotic BH3-only proteins such as BIM and PUMA are induced in two different models of scrapie, suggesting that other ER stress related pro-apoptotic pathways may modulate neuronal loss in infectious Prion-related disorders [110,111]. However, neither BAX deletion or BCL-2 overexpression affect infectious prion pathogenesis, suggesting that apoptosis is not essential for disease progression, and that it may be a downstream effect of neuronal dysfunction [110].

In addition to cell death with apoptotic features, autophagic vesicles are also observed in the brain of patients affected with PrD [112]. In another report using prion-infected neuronal cells, mild proteasome impairment resulted in the formation of large cytosolic perinuclear aggresomes that contained PrP^{Sc}, heat shock chaperone 70, ubiquitin, proteasome subunits and vimentin [113]. The role of autophagy in the progression of prion diseases is unknown, but the potential to promote clearance of misfolded protein aggregates may be a promising therapeutic approach [114].

Based on several *in vitro* and animal studies, different categories of drugs have been suggested as candidates for treatment of prion disease [115] [116]. Therapeutic interventions in human have included the use of antiviral drugs (*acyclovir*, *amantadine*, *vidarabine*), antimalarial (*quinacrina*), antifungal (*amphotericin*), anticoagulant (*pentosan polysulfate*), anticonvulsant (*levetiracetam*, *topiramate*, *phenytoin*) and analgesic drugs (*flupirtine*), together with antioxidants, vitamins and biologic response modifiers including *interferon* [117]. However, despite a marked *in vitro* "anti-prion" activity of some drugs tested, including quinacrina, a transient benefit or no effect on clinical symptoms are observed. Moreover, these trials did not appear to stop disease progression because no significant differences in survival following treatments were reported [117].

Chemical chaperones have been employed to reduce PrP^{C} to PrP^{Sc} conversion in mouse neuroblastoma cells. Exposure to compounds with chemical chaperone activity, including *glycerol*, *trimethylamine N-oxide dihydrate* (TMAO) and *dimethylsulfoxide* (DMSO), can influence PrP folding, stabilizing PrP^{C} and preventing prion replication [118]. Anilino-naphtalene compounds, which are used to label amyloid deposits, inhibit the aggregation of a PrP neurotoxic peptide, also binding to PrP^{Sc} [119]. *Doxycycline*, a semi-synthetic tetracycline, also showed an anti-prion activity, reverting the effects of PrP^{Sc} in a cellular model [120]. Since *doxycycline* is already approved for use in humans, future clinical trials are required to establish the efficacy of treating patients affected with TSEs. The discovery of novel chemical chaperones with low toxicity and higher permeability to the blood-brain barrier could lead to promising methods for treating this devastating disease

Beta-sheet breakers constitute a new class of drugs that were originally designed to specifically bind Alzeimer's disease amyloid-beta peptide and block or reverse the abnormal conformational change [121,122]. The effects of a 13-residue beta-sheet breaker peptide (iPrP13) were shown to be effective in reverting PrP^{Sc} to a biochemical and structural state similar to that of PrP^{C} .[122] More importantly, treatment of PrP^{Sc} with iPrP13 delayed the appearance of clinical symptoms and decreased infectivity by 90-95% in mice with experimental scrapie [122]. These peptides or their derivatives provide a useful tool to study the role of PrP conformation and may represent a novel therapeutic approach for prion-related disorders.

Cerebral Ischemia: The Use of Chemical Chaperones to Alleviate ER Stress and Increase Neuronal Survival

Clinically, cerebral ischemia often occurs as a result of heart attack or arterial occlusion leading to localized stroke. While factors causing ischemia can often be reversed and blood flow restored, cells that do not die from the initial anoxic insult face a recovery process that includes both avoiding damage from reactive oxygen species (ROS) produced during mitochondrial dysfunction and the resumption of protein synthesis in an unfavorable intracellular environment.

The UPR has been shown to be active both during early response to ischemic injury [123-125], and during the recovery process [126]. PERK mediated inhibition of protein synthesis minimizes the impact of ROS in damaged/oxidized proteins and other toxic events, During recovery, an increase in IRE1, XBP-1 mRNA splicing and ATF6 levels can be observed weeks after the initial event, indicating that ER stress is actively occurring in cells that do not succumb to apoptosis. Experimental evidence indicates that restoration of protein synthesis and removal of damaged, potentially toxic proteins by UPR mediated processes during the recovery period is essential to avoid additional neuronal loss. On the other hand, in models of hypoxia, the UPR has been shown to be an essential factor for cell survival [127-132].

Clinical studies have shown that treatment with edaravone, a free radical scavenger capable of removing ROS during ischemia, can result in improved recovery after acute stroke [133,134]. The relationship between reduction of ROS and improved recovery is most strongly correlated with suppression of mitochondrial dysfunction, but apoptosis is also initiated through caspase processing during periods of ER stress resulting from ROS-mediated protein misfolding [135]. While the effect of decreasing ROS has not been directly studied in terms of the UPR, correlative evidence indicates that agents capable of increasing ER protein folding capacity, such as 4-PBA, are similarly effective in reducing cell death [136]. Treatment of patients to both reduce oxidative stress and increase ER protein folding capabilities using chemical chaperones has potential to additively limit cellular damage and aid in recovery, therefore reducing cognitive and motor impairment resulting from ischemic trauma. As several chemical chaperones have now been approved for clinical use, combinatory clinical trials are now necessary to validate evidence from animal studies in human patients.

III. Concluding Remarks

The UPR is an essential pathway that controls adaptive processes to stress, which results in global changes of ER function including ERAD, ER/Golgi biogenesis, protein folding, translocation into the ER, and autophagy [137]. UPR genes such as XBP-1, ATF4 and ATF6 enforce changes in cellular structure and function consistent with the requirements of the UPR to maintain proper function of professional secretory cells, but the exact role of the UPR in the CNS is mostly unknown. A genetic link is observed between an XBP-1 promoter polymorphism and the occurrence of bipolar disorders [138], schizophrenia [139] and certain personality types in the Japanese population [140]. As discussed here, extensive studies indicate a strong association between accumulation of misfolded proteins and ER stress induction in several important neurodegenerative conditions [29,141]. Direct evidence indicating that perturbation of ER function could result in neurodegeneration came from the characterization of the Woozy mutant mouse, where disruption of a BiP co-chaperone triggers neuronal dysfunction associated with spontaneous protein aggregation in the brain [142]. Similarly, recent reports indicate that autophagy is critical for the maintenance of neuronal homoeostasis and has a role in the basal elimination of misfolded, ubiquitinated proteins. Brain specific knockout animals for essential autophagic-related genes develop spontaneous neurodegeneration with pathological features similar to Alzheimer's and Parkinson's disease [56,57]. These data suggest that the continuous clearance of cellular proteins in the brain through basal autophagy is required to prevent abnormal accumulation, therefore acting as a neuroprotective mechanism. However, most of the evidence supporting the involvement of ER stress in neurodegeneration is correlative, and manipulation of the UPR *in vivo* was required to define the actual contribution of the pathway to the disease process.

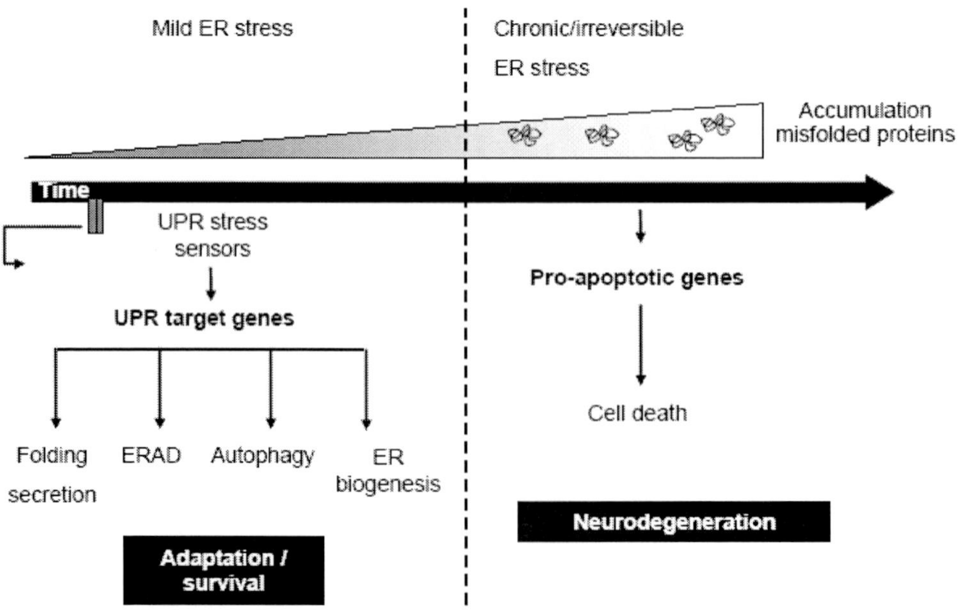

Figure 3. Protein misfolding and neurodegeneration. The accumulation and deposition of specific misfolded proteins is common in many neurodegenerative diseases. Activation of various components of the UPR is observed in several protein misfolding disorders affecting the nervous system. The early UPR response activates survival pathways to reestablish cellular homeostasis and inhibit apoptosis. These responses include increased protein folding through the upregulation of chaperones, increased protein quality control, increased ER biogenesis, and enhancement of autophagy. Under chronic or irreversible ER damage the UPR ultimately initiates apoptosis to eliminate damaged cells, contributing to neurodegenerative processes of disease.

Strong correlations have been described between the misfolding and aggregation of an underlying protein and the occurrence of ER stress in neurodegenerative conditions. Additionally, increasing reports describe activation of the UPR in human post-mortem samples from patients affected with diverse types of protein misfolding disorders, suggesting a role as a general cellular response to neurodegeneration (Table 1). Despite this evidence, little causal relationship is available to link the UPR and ER stress to neurological disorders *in vivo*. Predictions for the role of ER stress in disease processes are not obvious because activation of the UPR may decrease neurodegeneration by increasing folding, protein quality control and autophagy. However, extensive or chronic ER stress may result in irreversible neuronal damage and apoptosis (Figure 3). Promising results have been obtained with pharmacological strategies to target ER stress in a disease context. Small molecules, including chemical chaperones and autophagy activators, have been shown to have protective effects against neurodegeneration in certain disease models *in vivo*, decreasing protein misfolding and aggregation. Genetic manipulation of UPR components *in vivo* will ultimately identify the role of these processes in neurodegeneration.

Acknowledgments

This work was supported by FONDECYT no. 1070444, FONDAP grant no.15010006, High Q Foundation, Michael J Fox Foundation for Parkinson's Research, The Muscular Dystrophy Association (CH), FONCECYT no. 3085017 (SM), and CONYCIT Ph.D fellowship (MJ and MT).

References

[1] R. R. Kopito and D. Ron, Conformational disease, *Nat. Cell Biol., 2* (2000) E207-E209.

[2] R. V. Rao and D. E. Bredesen, Misfolded proteins, endoplasmic reticulum stress and neurodegeneration, *Curr. Opin. Cell Biol., 16* (2004) 653-662.

[3] D. J. Selkoe, Folding proteins in fatal ways, *Nature, 426* (2003) 900-904.

[4] J. P. Taylor, J. Hardy, and K. H. Fischbeck, Toxic proteins in neurodegenerative disease, *Science, 296* (2002) 1991-1995.

[5] D. Ron and P. Walter, Signal integration in the endoplasmic reticulum unfolded protein response, *Nat. Rev. Mol. Cell Biol., 8* (2007) 519-529.

[6] M. Schroder and R. J. Kaufman, The mammalian unfolded protein response, *Annu. Rev. Biochem., 74* (2005) 739-789.

[7] C. A. Hetz, V. Torres, and A. F. Quest, Beyond apoptosis: nonapoptotic cell death in physiology and disease, *Biochem. Cell Biol., 83* (2005) 579-588.

[8] C. Hetz and L. Glimcher, XBP-1 and the UPRosome: Mastering Secretory Cell Function, *Current Immunology Reviews. 4* (2008):1-10.

[9] H. P. Harding, Y. Zhang, and D. Ron, Protein translation and folding are coupled by an endoplasmic-reticulum-resident kinase, *Nature, 397* (1999) 271-274.

[10] X. Chen, J. Shen, and R. Prywes, The luminal domain of ATF6 senses endoplasmic reticulum (ER) stress and causes translocation of ATF6 from the ER to the Golgi, *J. Biol. Chem., 277* (2002) 13045-13052.

[11] K. Haze, H. Yoshida, H. Yanagi, T. Yura, and K. Mori, Mammalian transcription factor ATF6 is synthesized as a transmembrane protein and activated by proteolysis in response to endoplasmic reticulum stress, *Mol. Biol. Cell, 10* (1999) 3787-3799.

[12] J. Ye, R. B. Rawson, R. Komuro, X. Chen, U. P. Dave, R. Prywes, M. S. Brown, and J. L. Goldstein, ER stress induces cleavage of membrane-bound ATF6 by the same proteases that process SREBPs, *Mol. Cell, 6* (2000) 1355-1364.

[13] M. Calfon, H. Zeng, F. Urano, J. H. Till, S. R. Hubbard, H. P. Harding, S. G. Clark, and D. Ron, IRE1 couples endoplasmic reticulum load to secretory capacity by processing the XBP-1 mRNA, *Nature, 415* (2002) 92-96.

[14] K. Lee, W. Tirasophon, X. Shen, M. Michalak, R. Prywes, T. Okada, H. Yoshida, K. Mori, and R. J. Kaufman, IRE1-mediated unconventional mRNA splicing and S2P-mediated ATF6 cleavage merge to regulate XBP1 in signaling the unfolded protein response, *Genes Dev., 16* (2002) 452-466.

[15] H. Yoshida, T. Matsui, A. Yamamoto, T. Okada, and K. Mori, XBP1 mRNA is induced by ATF6 and spliced by IRE1 in response to ER stress to produce a highly active transcription factor, *Cell, 107* (2001) 881-891.

[16] C. Hetz and L. Glimcher, The daily job of night killers: alternative roles of the BCL-2 family in organelle physiology, *Trends Cell Biol.*, (2007).

[17] C. Hetz, P. Bernasconi, J. Fisher, A. H. Lee, M. C. Bassik, B. Antonsson, G. S. Brandt, N. N. Iwakoshi, A. Schinzel, L. H. Glimcher, and S. J. Korsmeyer, Proapoptotic BAX and BAK modulate the unfolded protein response by a direct interaction with IRE1alpha, *Science, 312* (2006) 572-576.

[18] R. R. Kopito and D. Ron, Conformational disease, Nat. Cell Biol., 2 (2000) E207-E209.

[19] D. J. Selkoe, Folding proteins in fatal ways, *Nature, 426* (2003) 900-904.

[20] J. P. Taylor, J. Hardy, and K. H. Fischbeck, Toxic proteins in neurodegenerative disease, *Science, 296* (2002) 1991-1995.

[21] J. J. Hoozemans, E. S. van Haastert, P. Eikelenboom, R. A. de Vos, J. M. Rozemuller, and W. Scheper, Activation of the unfolded protein response in Parkinson's disease, *Biochem. Biophys. Res. Commun., 354* (2007) 707-711.

[22] T. Uehara, T. Nakamura, D. Yao, Z. Q. Shi, Z. Gu, Y. Ma, E. Masliah, Y. Nomura, and S. A. Lipton, S-nitrosylated protein-disulphide isomerase links protein misfolding to neurodegeneration, *Nature, 441* (2006) 513-517.

[23] K. J. Conn, W. Gao, A. McKee, M. S. Lan, M. D. Ullman, P. B. Eisenhauer, R. E. Fine, and J. M. Wells, Identification of the protein disulfide isomerase family member PDIp in experimental Parkinson's disease and Lewy body pathology, *Brain Res., 1022* (2004) 164-172.

[24] E. V. Ilieva, V. Ayala, M. Jove, E. Dalfo, D. Cacabelos, M. Povedano, M. J. Bellmunt, I. Ferrer, R. Pamplona, and M. Portero-Otin, Oxidative and endoplasmic reticulum stress interplay in sporadic amyotrophic lateral sclerosis, *Brain, 130* (2007) 3111-3123.

[25] J. D. Atkin, M. A. Farg, A. K. Walker, C. McLean, D. Tomas, and M. K. Horne, Endoplasmic reticulum stress and induction of the unfolded protein response in human sporadic amyotrophic lateral sclerosis, *Neurobiol. Dis.*, (2008).

[26] U. Unterberger, R. Hoftberger, E. Gelpi, H. Flicker, H. Budka, and T. Voigtlander, Endoplasmic reticulum stress features are prominent in Alzheimer disease but not in prion diseases in vivo, *J. Neuropathol. Exp. Neurol., 65* (2006) 348-357.

[27] C. Hetz, M. Russelakis-Carneiro, K. Maundrell, J. Castilla, and C. Soto, Caspase-12 and endoplasmic reticulum stress mediate neurotoxicity of pathological prion protein, *EMBO J., 22* (2003) 5435-5445.

[28] B. C. Yoo, K. Krapfenbauer, N. Cairns, G. Belay, M. Bajo, and G. Lubec, Overexpressed protein disulfide isomerase in brains of patients with sporadic Creutzfeldt-Jakob disease, *Neurosci. Lett., 334* (2002) 196-200.

[29] D. Lindholm, H. Wootz, and L. Korhonen, ER stress and neurodegenerative diseases, *Cell Death. Differ., 13* (2006) 385-392.

[30] S. Matus, F. Lisbona, M. Torres, C. Leon, P. Thielen, and C. Hetz, The stress rheostat: an interplay between the unfolded protein response (UPR) and autophagy in neurodegeneration, *Curr. Mol. Med., 8* (2008) 157-172.

[31] HDCRG, A novel gene containing a trinucleotide repeat that is expanded and unstable on Huntington's disease chromosomes. The Huntington's Disease Collaborative Research Group, *Cell, 72* (1993) 971-983.

[32] B. Kremer, P. Goldberg, S. E. Andrew, J. Theilmann, H. Telenius, J. Zeisler, F. Squitieri, B. Lin, A. Bassett, E. Almqvist, and ., A worldwide study of the Huntington's disease mutation. The sensitivity and specificity of measuring CAG repeats, *N. Engl. J. Med., 330* (1994) 1401-1406.

[33] D. C. Rubinsztein, Lessons from animal models of Huntington's disease, *Trends Genet., 18* (2002) 202-209.

[34] A. S. Hackam, R. Singaraja, C. L. Wellington, M. Metzler, K. McCutcheon, T. Zhang, M. Kalchman, and M. R. Hayden, The influence of huntingtin protein size on nuclear localization and cellular toxicity, *J. Cell Biol., 141* (1998) 1097-1105.

[35] A. Lunkes and J. L. Mandel, A cellular model that recapitulates major pathogenic steps of Huntington's disease, *Hum. Mol. Genet., 7* (1998) 1355-1361.

[36] A. Wyttenbach, J. Carmichael, J. Swartz, R. A. Furlong, Y. Narain, Rankin .J, and D. C. Rubinsztein, Effects of heat shock, heat shock protein 40 (HDJ-2), and proteasome inhibition on protein aggregation in cellular models of Huntington's disease, *Proc. Natl. Acad. Sci. USA, 97* (2000) 2898-2903.

[37] F. C. Nucifora, Jr., M. Sasaki, M. F. Peters, H. Huang, J. K. Cooper, M. Yamada, H. Takahashi, S. Tsuji, J. Troncoso, V. L. Dawson, T. M. Dawson, and C. A. Ross, Interference by huntingtin and atrophin-1 with cbp-mediated transcription leading to cellular toxicity, *Science, 291* (2001) 2423-2428.

[38] B. Ravikumar and D. C. Rubinsztein, Role of autophagy in the clearance of mutant huntingtin: a step towards therapy?, *Mol. Aspects Med., 27* (2006) 520-527.

[39] T. Momoi, Conformational diseases and ER stress-mediated cell death: apoptotic cell death and autophagic cell death, *Curr. Mol. Med., 6* (2006) 111-118.

[40] R. R. Kopito, Aggresomes, inclusion bodies and protein aggregation, *Trends Cell Biol., 10* (2000) 524-530.

[41] K. Omi, N. S. Hachiya, K. Tokunaga, and K. Kaneko, siRNA-mediated inhibition of endogenous Huntington disease gene expression induces an aberrant configuration of the ER network in vitro, *Biochem. Biophys. Res. Commun., 338* (2005) 1229-1235.

[42] E. Rockabrand, N. Slepko, A. Pantalone, V. N. Nukala, A. Kazantsev, J. L. Marsh, P. G. Sullivan, J. S. Steffan, S. L. Sensi, and L. M. Thompson, The first 17 amino acids of Huntingtin modulate its sub-cellular localization, aggregation and effects on calcium homeostasis, *Hum. Mol. Genet., 16* (2007) 61-77.

[43] E. Rousseau, B. Dehay, L. Ben-Haiem, Y. Trottier, M. Morange, and A. Bertolotti, Targeting expression of expanded polyglutamine proteins to the endoplasmic reticulum or mitochondria prevents their aggregation, *Proc. Natl. Acad. Sci. U. S. A, 101* (2004) 9648-9653.

[44] Y. Kouroku, E. Fujita, I. Tanida, T. Ueno, A. Isoai, H. Kumagai, S. Ogawa, R. J. Kaufman, E. Kominami, and T. Momoi, ER stress (PERK/eIF2alpha phosphorylation) mediates the polyglutamine-induced LC3 conversion, an essential step for autophagy formation, *Cell Death. Differ.*, (2006).

[45] T. Nakagawa, H. Zhu, N. Morishima, E. Li, J. Xu, B. A. Yankner, and J. Yuan, Caspase-12 mediates endoplasmic-reticulum-specific apoptosis and cytotoxicity by amyloid-beta, *Nature, 403* (2000) 98-103.

[46] H. Nishitoh, A. Matsuzawa, K. Tobiume, K. Saegusa, K. Takeda, K. Inoue, S. Hori, A. Kakizuka, and H. Ichijo, ASK1 is essential for endoplasmic reticulum stress-induced neuronal cell death triggered by expanded polyglutamine repeats, *Genes Dev., 16* (2002) 1345-1355.

[47] N. R. Jana, M. Tanaka, G. Wang, and N. Nukina, Polyglutamine length-dependent interaction of Hsp40 and Hsp70 family chaperones with truncated N-terminal huntingtin: their role in suppression of aggregation and cellular toxicity, *Hum. Mol. Genet., 9* (2000) 2009-2018.

[48] M. C. Maiuri, E. Zalckvar, A. Kimchi, and G. Kroemer, Self-eating and self-killing: crosstalk between autophagy and apoptosis, *Nat. Rev. Mol. Cell Biol., 8* (2007) 741-752.

[49] U. Ozcan, E. Yilmaz, L. Ozcan, M. Furuhashi, E. Vaillancourt, R. O. Smith, C. Z. Gorgun, and G. S. Hotamisligil, Chemical chaperones reduce ER stress and restore glucose homeostasis in a mouse model of type 2 diabetes, *Science, 313* (2006) 1137-1140.

[50] W. J. Welch and C. R. Brown, Influence of molecular and chemical chaperones on protein folding, Cell Stress. *Chaperones., 1* (1996) 109-115.

[51] Q. Xie, V. I. Khaoustov, C. C. Chung, J. Sohn, B. Krishnan, D. E. Lewis, and B. Yoffe, Effect of tauroursodeoxycholic acid on endoplasmic reticulum stress-induced caspase-12 activation, *Hepatology, 36* (2002) 592-601.

[52] C. D. Keene, C. M. Rodrigues, T. Eich, M. S. Chhabra, C. J. Steer, and W. C. Low, Tauroursodeoxycholic acid, a bile acid, is neuroprotective in a transgenic animal model of Huntington's disease, *Proc. Natl. Acad. Sci. U. S. A, 99* (2002) 10671-10676.

[53] R. J. Ferrante, J. K. Kubilus, J. Lee, H. Ryu, A. Beesen, B. Zucker, K. Smith, N. W. Kowall, R. R. Ratan, R. Luthi-Carter, and S. M. Hersch, Histone deacetylase inhibition by sodium butyrate chemotherapy ameliorates the neurodegenerative phenotype in Huntington's disease mice, *J. Neurosci., 23* (2003) 9418-9427.

[54] M. Hoyer-Hansen and M. Jaattela, Connecting endoplasmic reticulum stress to autophagy by unfolded protein response and calcium, *Cell Death. Differ.*, (2007).

[55] M. Komatsu, S. Waguri, T. Chiba, S. Murata, J. Iwata, I. Tanida, T. Ueno, M. Koike, Y. Uchiyama, E. Kominami, and K. Tanaka, Loss of autophagy in the central nervous system causes neurodegeneration in mice, *Nature, 441* (2006) 880-884.

[56] T. Hara, K. Nakamura, M. Matsui, A. Yamamoto, Y. Nakahara, R. Suzuki-Migishima, M. Yokoyama, K. Mishima, I. Saito, H. Okano, and N. Mizushima, Suppression of basal autophagy in neural cells causes neurodegenerative disease in mice, *Nature, 441* (2006) 885-889.

[57] B. Ravikumar and D. C. Rubinsztein, Can autophagy protect against neurodegeneration caused by aggregate-prone proteins?, *Neuroreport, 15* (2004) 2443-2445.

[58] R. A. Floto, S. Sarkar, E. O. Perlstein, B. Kampmann, S. L. Schreiber, and D. C. Rubinsztein, Small molecule enhancers of rapamycin-induced TOR inhibition promote autophagy, reduce toxicity in Huntington's disease models and enhance killing of mycobacteria by macrophages, *Autophagy., 3* (2007) 620-622.

[59] S. Sarkar, J. E. Davies, Z. Huang, A. Tunnacliffe, and D. C. Rubinsztein, Trehalose, a novel mTOR-independent autophagy enhancer, accelerates the clearance of mutant huntingtin and alpha-synuclein, *J. Biol. Chem., 282* (2007) 5641-5652.

[60] S. Sarkar, E. O. Perlstein, S. Imarisio, S. Pineau, A. Cordenier, R. L. Maglathlin, J. A. Webster, T. A. Lewis, C. J. O'Kane, S. L. Schreiber, and D. C. Rubinsztein, Small molecules enhance autophagy and reduce toxicity in Huntington's disease models, *Nat. Chem. Biol., 3* (2007) 331-338.

[61] S. Sarkar, E. O. Perlstein, S. Imarisio, S. Pineau, A. Cordenier, R. L. Maglathlin, J. A. Webster, T. A. Lewis, C. J. O'Kane, S. L. Schreiber, and D. C. Rubinsztein, Small molecules enhance autophagy and reduce toxicity in Huntington's disease models, *Nat. Chem. Biol., 3* (2007) 331-338.

[62] P. Pasinelli and R. H. Brown, Molecular biology of amyotrophic lateral sclerosis: insights from genetics, *Nat. Rev. Neurosci., 7* (2006) 710-723.

[63] S. Boillee, V. C. Vande, and D. W. Cleveland, ALS: a disease of motor neurons and their nonneuronal neighbors, *Neuron, 52* (2006) 39-59.

[64] B. J. Turner and J. D. Atkin, ER Stress and UPR in Familial Amyotrophic Lateral Sclerosis, *Curr. Mol. Med., 6* (2006) 79-86.

[65] A. S. Vlug, E. Teuling, E. D. Haasdijk, P. French, C. C. Hoogenraad, and D. Jaarsma, ATF3 expression precedes death of spinal motoneurons in amyotrophic lateral sclerosis-SOD1 transgenic mice and correlates with c-Jun phosphorylation, CHOP expression, somato-dendritic ubiquitination and Golgi fragmentation, *Eur. J. Neurosci., 22* (2005) 1881-1894.

[66] T. Nagata, H. Ilieva, T. Murakami, M. Shiote, H. Narai, Y. Ohta, T. Hayashi, M. Shoji, and K. Abe, Increased ER stress during motor neuron degeneration in a transgenic mouse model of amyotrophic lateral sclerosis, *Neurol. Res., 29* (2007) 767-771.

[67] H. Kikuchi, G. Almer, S. Yamashita, C. Guegan, M. Nagai, Z. Xu, A. A. Sosunov, G. M. McKhann, and S. Przedborski, Spinal cord endoplasmic reticulum stress associated with a microsomal accumulation of mutant superoxide dismutase-1 in an ALS model, *Proc. Natl. Acad. Sci. U. S. A, 103* (2006) 6025-6030.

[68] H. Wootz, I. Hansson, L. Korhonen, U. Napankangas, and D. Lindholm, Caspase-12 cleavage and increased oxidative stress during motoneuron degeneration in transgenic mouse model of ALS, *Biochem. Biophys. Res. Commun., 322* (2004) 281-286.

[69] J. D. Atkin, M. A. Farg, B. J. Turner, D. Tomas, J. A. Lysaght, J. Nunan, A. Rembach, P. Nagley, P. M. Beart, S. S. Cheema, and M. K. Horne, Induction of the unfolded protein response in familial amyotrophic lateral sclerosis and association of protein-disulfide isomerase with superoxide dismutase 1, *J. Biol. Chem., 281* (2006) 30152-30165.

[70] M. Urushitani, A. Sik, T. Sakurai, N. Nukina, R. Takahashi, and J. P. Julien, Chromogranin-mediated secretion of mutant superoxide dismutase proteins linked to amyotrophic lateral sclerosis, *Nat. Neurosci., 9* (2006) 108-118.

[71] H. Wootz, I. Hansson, L. Korhonen, and D. Lindholm, XIAP decreases caspase-12 cleavage and calpain activity in spinal cord of ALS transgenic mice, *Exp. Cell Res.,* (2006).

[72] D. Kieran, I. Woods, A. Villunger, A. Strasser, and J. H. Prehn, Deletion of the BH3-only protein puma protects motoneurons from ER stress-induced apoptosis and delays motoneuron loss in ALS mice, *Proc. Natl. Acad. Sci. U. S. A, 104* (2007) 20606-20611.

[73] N. Morishima, K. Nakanishi, K. Tsuchiya, T. Shibata, and E. Seiwa, Translocation of Bim to the endoplasmic reticulum (ER) mediates ER stress signaling for activation of caspase-12 during ER stress-induced apoptosis, *J. Biol. Chem., 279* (2004) 50375-50381.

[74] H. Puthalakath, L. A. O'Reilly, P. Gunn, L. Lee, P. N. Kelly, N. D. Huntington, P. D. Hughes, E. M. Michalak, J. Kimm-Breschkin, N. Motoyama, T. Gotoh, S. Akira, P. Bouillet, and A. Strasser, ER Stress Triggers Apoptosis by Activating BH3-Only Protein Bim, *Cell, 129* (2007) 1337-1349.

[75] C. Hetz, P. Thielen, J. Fisher, P. Pasinelli, R. H. Brown, S. Korsmeyer, and L. Glimcher, The proapoptotic BCL-2 family member BIM mediates motoneuron loss in a model of amyotrophic lateral sclerosis, *Cell Death. Differ.*, (2007).

[76] T. W. Gould, R. R. Buss, S. Vinsant, D. Prevette, W. Sun, C. M. Knudson, C. E. Milligan, and R. W. Oppenheim, Complete dissociation of motor neuron death from motor dysfunction by Bax deletion in a mouse model of ALS, *J. Neurosci., 26* (2006) 8774-8786.

[77] S. Vukosavic, L. Stefanis, V. Jackson-Lewis, C. Guegan, N. Romero, C. Chen, M. Dubois-Dauphin, and S. Przedborski, Delaying caspase activation by Bcl-2: A clue to disease retardation in a transgenic mouse model of amyotrophic lateral sclerosis, *J. Neurosci., 20* (2000) 9119-9125.

[78] M. Boyce, K. F. Bryant, C. Jousse, K. Long, H. P. Harding, D. Scheuner, R. J. Kaufman, D. Ma, D. M. Coen, D. Ron, and J. Yuan, A selective inhibitor of eIF2alpha dephosphorylation protects cells from ER stress, *Science, 307* (2005) 935-939.

[79] Y. K. Oh, K. S. Shin, J. Yuan, and S. J. Kang, Superoxide dismutase 1 mutants related to amyotrophic lateral sclerosis induce endoplasmic stress in neuro2a cells, *J. Neurochem., 104* (2008) 993-1005.

[80] C. Raoul, T. bbas-Terki, J. C. Bensadoun, S. Guillot, G. Haase, J. Szulc, C. E. Henderson, and P. Aebischer, Lentiviral-mediated silencing of SOD1 through RNA interference retards disease onset and progression in a mouse model of ALS, *Nat. Med., 11* (2005) 423-428.

[81] M. Azzouz, A. Hottinger, J. C. Paterna, A. D. Zurn, P. Aebischer, and H. Bueler, Increased motoneuron survival and improved neuromuscular function in transgenic ALS mice after intraspinal injection of an adeno-associated virus encoding Bcl-2, *Hum. Mol. Genet., 9* (2000) 803-811.

[82] D. Kirik and A. Bjorklund, Modeling CNS neurodegeneration by overexpression of disease-causing proteins using viral vectors, *Trends Neurosci., 26* (2003) 386-392.

[83] M. Azzouz, S. M. Kingsman, and N. D. Mazarakis, Lentiviral vectors for treating and modeling human CNS disorders, *J. Gene Med., 6* (2004) 951-962.

[84] T. Kabuta, Y. Suzuki, and K. Wada, Degradation of amyotrophic lateral sclerosis-linked mutant Cu,Zn-superoxide dismutase proteins by macroautophagy and the proteasome, *J. Biol. Chem., 281* (2006) 30524-30533.

[85] N. Morimoto, M. Nagai, Y. Ohta, K. Miyazaki, T. Kurata, M. Morimoto, T. Murakami, Y. Takehisa, Y. Ikeda, T. Kamiya, and K. Abe, Increased autophagy in transgenic mice with a G93A mutant SOD1 gene, *Brain Res., 1167* (2007) 112-117.

[86] F. Fornai, P. Longone, L. Cafaro, O. Kastsiuchenka, M. Ferrucci, M. L. Manca, G. Lazzeri, A. Spalloni, N. Bellio, P. Lenzi, N. Modugno, G. Siciliano, C. Isidoro, L. Murri, S. Ruggieri, and A. Paparelli, Lithium delays progression of amyotrophic lateral sclerosis, *Proc. Natl. Acad. Sci. U. S. A, 105* (2008) 2052-2057.

[87] S. Sarkar, R. A. Floto, Z. Berger, S. Imarisio, A. Cordenier, M. Pasco, L. J. Cook, and D. C. Rubinsztein, Lithium induces autophagy by inhibiting inositol monophosphatase, *J. Cell Biol., 170* (2005) 1101-1111.

[88] S. B. Prusiner, Prions, *Proc. Natl. Acad. Sci. USA, 95* (1998) 13363-13383.

[89] J. Collinge, Prion diseases of humans and animals: their causes and molecular basis, *Annu. Rev. Neurosci., 24* (2001) 519-550.

[90] H. Budka, A. Aguzzi, P. Brown, J. M. Brucher, O. Bugiani, F. Gullotta, M. Haltia, J. J. Hauw, J. W. Ironside, K. Jellinger, H. A. Kretzschmar, P. L. Lantos, C. Masullo, W. Schlote, J. Tateishi, and R. O. Weller, Neuropathological diagnostic criteria for Creutzfeldt-Jakob disease (CJD) and other human spongiform encephalopathies (Prion diseases), *Brain Pathol., 5* (1995) 459-466.

[91] K. M. Pan, M. Baldwin, J. Nguyen, M. Gasset, A. Serban, D. Groth, I. Mehlhorn, Z. Huang, R. J. Fletterick, F. E. Cohen, and ., Conversion of alpha-helices into beta-sheets features in the formation of the scrapie prion proteins, *Proc. Natl. Acad. Sci. U. S. A, 90* (1993) 10962-10966.

[92] M. Vey, S. Pilkuhn, H. Wille, R. Nixon, S. J. DeArmond, E. J. Smart, R. G. Anderson, A. Taraboulos, and S. B. Prusiner, Subcellular colocalization of the cellular and scrapie prion proteins in caveolae-like membranous domains, *Proc. Natl. Acad. Sci. U. S A, 93* (1996) 14945-14949.

[93] C. Hetz, K. Maundrell, and C. Soto, Is loss of function of the prion protein the cause of prion disorders?, *Trends Mol. Med., 9* (2003) 237-243.

[94] C. Hetz and C. Soto, Protein misfolding and disease: the case of prion disorders, *Cell Mol. Life Sci., 60* (2003) 133-143.

[95] R. Chiesa, B. Drisaldi, E. Quaglio, A. Migheli, P. Piccardo, B. Ghetti, and D. A. Harris, Accumulation of protease-resistant prion protein (PrP) and apoptosis of cerebellar granule cells in transgenic mice expressing a PrP insertional mutation, *Proc. Natl. Acad. Sci. U. S. A, 97* (2000) 5574-5579.

[96] J. Ma, R. Wollmann, and S. Lindquist, Neurotoxicity and neurodegeneration when PrP accumulates in the cytosol, *Science, 298* (2002) 1781-1785.

[97] S. Supattapone, E. Bouzamondo, H. L. Ball, H. Wille, H. O. Nguyen, F. E. Cohen, S. J. DeArmond, S. B. Prusiner, and M. Scott, A protease-resistant 61-residue prion peptide causes neurodegeneration in transgenic mice, *Mol. Cell Biol., 21* (2001) 2608-2616.

[98] S. Supattapone, P. Bosque, T. Muramoto, H. Wille, C. Aagaard, D. Peretz, H. O. Nguyen, C. Heinrich, M. Torchia, J. Safar, F. E. Cohen, S. J. DeArmond, S. B. Prusiner, and M. Scott, Prion protein of 106 residues creates an artifical transmission barrier for prion replication in transgenic mice, *Cell, 96* (1999) 869-878.

[99] C. Hetz, A. H. Lee, D. Gonzalez-Romero, P. Thielen, J. Castilla, C. Soto, and L. H. Glimcher, Unfolded protein response transcription factor XBP-1 does not influence prion replication or pathogenesis, *Proc. Natl. Acad. Sci. U. S. A, 105*(2008): 757-62.

[100] A. D. Steele, C. Hetz, C. Yi, W. Jackson, A. Borkowski, J. Yuan, R. Wollmann, and S. Lindquist, Prion disease in caspase-12 knockout mice, *Proc. Natl. Acad. Sci. (USA)*, (2008).

[101] A. R. Brown, S. Rebus, C. S. McKimmie, K. Robertson, A. Williams, and J. K. Fazakerley, Gene expression profiling of the preclinical scrapie-infected hippocampus, *Biochem. Biophys. Res. Commun., 334* (2005) 86-95.

[102] C. Hetz, M. Russelakis-Carneiro, S. Walchli, S. Carboni, E. Vial-Knecht, K. Maundrell, J. Castilla, and C. Soto, The disulfide isomerase Grp58 is a protective factor against prion neurotoxicity, *J. Neurosci., 25* (2005) 2793-2802.

[103] C. Hetz, J. Castilla, and C. Soto, Perturbation of endoplasmic reticulum homeostasis facilitates prion replication, *J. Biol. Chem., 282* (2007) 12725-12733.

[104] A. Orsi, L. Fioriti, R. Chiesa, and R. Sitia, Conditions of endoplasmic reticulum stress favor the accumulation of cytosolic prion protein, *J. Biol. Chem., 281* (2006) 30431-30438.

[105] J. Apodaca, I. Kim, and H. Rao, Cellular tolerance of prion protein PrP in yeast involves proteolysis and the unfolded protein response, *Biochem. Biophys. Res. Commun., 347* (2006) 319-326.

[106] C. Hetz, A. H. Lee, D. Gonzalez-Romero, P. Thielen, J. Castilla, C. Soto, and L. H. Glimcher, Unfolded protein response transcription factor XBP-1 does not influence prion replication or pathogenesis, *Proc. Natl. Acad. Sci. U. S. A*, (2008).

[107] Steele A., Hetz C., Yi C., Jackson W., Borkowski A., Yuan J., Wollmann R., and Lindquist S. Prion pathogenesis is independent of Caspase-12. *Prion. 4* (2007):1-5.

[108] A. D. Steele, O. D. King, W. S. Jackson, C. A. Hetz, A. W. Borkowski, P. Thielen, R. Wollmann, and S. Lindquist, Diminishing Apoptosis by Deletion of Bax or

Overexpression of Bcl-2 Does Not Protect against Infectious Prion Toxicity In Vivo, J. *Neurosci., 27* (2007) 13022-13027.

[109] B. Sikorska, P. P. Liberski, P. Giraud, N. Kopp, and P. Brown, Autophagy is a part of ultrastructural synaptic pathology in Creutzfeldt-Jakob disease: a brain biopsy study 69, *Int. J. Biochem. Cell Biol., 36* (2004) 2563-2573.

[110] M. Kristiansen, M. J. Messenger, P. C. Klohn, S. Brandner, J. D. Wadsworth, J. Collinge, and S. J. Tabrizi, Disease-related prion protein forms aggresomes in neuronal cells leading to caspase-activation and apoptosis, *J. Biol. Chem.*, (2005).

[111] D. C. Rubinsztein, J. E. Gestwicki, L. O. Murphy, and D. J. Klionsky, Potential therapeutic applications of autophagy 71, *Nat. Rev. Drug Discov., 6* (2007) 304-312.

[112] C. Korth, B. C. H. May, F. E. Cohen, and S. B. Prusiner, Acridine and phenothiazine derivatives as pharmacotherapeutics for prion disease, *Proc. Natl. Acad. Sci. USA, 98* (2001) 9836-9841.

[113] C. Korth and P. J. Peters, Emerging pharmacotherapies for Creutzfeldt-Jakob disease, *Arch. Neurol., 63* (2006) 497-501.

[114] L. A. Stewart, L. H. Rydzewska, G. F. Keogh, and R. S. Knight, Systematic review of therapeutic interventions in human prion disease, *Neurol., 70* (2008) 1272-1281.

[115] J. Tatzelt, S. B. Prusiner, and W. J. Welch, Chemical chaperones interfere with the formation of scrapie prion protein, *EMBO J., 15* (1996) 6363-6373.

[116] Y. Cordeiro, L. M. Lima, M. P. Gomes, D. Foguel, and J. L. Silva, Modulation of prion protein oligomerization, aggregation, and beta-sheet conversion by 4,4'-dianilino-1,1'-binaphthyl-5,5'-sulfonate (bis-ANS), *J. Biol. Chem., 279* (2004) 5346-5352.

[117] Y. Gu and N. Singh, Doxycycline and protein folding agents rescue the abnormal phenotype of familial CJD H187R in a cell model, *Brain Res. Mol. Brain Res., 123* (2004) 37-44.

[118] C. Soto, Unfolding the role of protein misfolding in neurodegenerative diseases, *Nat. Rev. Neurosci., 4* (2003) 49-60.

[119] C. Soto, R. J. Kascsak, G. P. Saborio, P. Aucouturier, T. Wisniewski, F. Prelli, R. Kascsak, E. Mendez, D. A. Harris, J. Ironside, F. Tagliavini, R. I. Carp, and B. Frangione, Reversion of prion protein conformational changes by synthetic ß-sheet breaker peptides, *Lancet, 355* (2000) 192-197.

[120] W. Paschen, Endoplasmic reticulum: a primary target in various acute disorders and degenerative diseases of the brain, *Cell Calcium, 34* (2003) 365-383.

[121] W. Paschen, C. Aufenberg, S. Hotop, and T. Mengesdorf, Transient cerebral ischemia activates processing of xbpl messenger RNA indicative of endoplasmic reticulum stress, *J. Cereb. Blood Flow Metab, 23* (2003) 449-461.

[122] R. Kumar, G. S. Krause, H. Yoshida, K. Mori, and D. J. DeGracia, Dysfunction of the unfolded protein response during global brain ischemia and reperfusion, *J. Cereb. Blood Flow Metab, 23* (2003) 462-471.

[123] A. Rissanen, J. Sivenius, and J. Jolkkonen, Prolonged bihemispheric alterations in unfolded protein response related gene expression after experimental stroke, *Brain Res., 1087* (2006) 60-66.

[124] C. Koumenis, M. Bi, J. Ye, D. Feldman, and A. C. Koong, Hypoxia and the unfolded protein response, *Methods Enzymol., 435* (2007) 275-293.

[125] C. Koumenis, ER stress, hypoxia tolerance and tumor progression, *Curr. Mol. Med., 6* (2006) 55-69.
[126] D. R. Fels and C. Koumenis, The PERK/eIF2alpha/ATF4 module of the UPR in hypoxia resistance and tumor growth, *Cancer Biol. Ther., 5* (2006) 723-728.
[127] M. Bi, C. Naczki, M. Koritzinsky, D. Fels, J. Blais, N. Hu, H. Harding, I. Novoa, M. Varia, J. Raleigh, D. Scheuner, R. J. Kaufman, J. Bell, D. Ron, B. G. Wouters, and C. Koumenis, ER stress-regulated translation increases tolerance to extreme hypoxia and promotes tumor growth, *EMBO J., 24* (2005) 3470-3481.
[128] L. Romero-Ramirez, H. Cao, D. Nelson, E. Hammond, A. H. Lee, H. Yoshida, K. Mori, L. H. Glimcher, N. C. Denko, A. J. Giaccia, Q. T. Le, and A. C. Koong, XBP1 is essential for survival under hypoxic conditions and is required for tumor growth, *Cancer Res., 64* (2004) 5943-5947.
[129] D. E. Feldman, V. Chauhan, and A. C. Koong, The unfolded protein response: a novel component of the hypoxic stress response in tumors, *Mol. Cancer Res., 3* (2005) 597-605.
[130] K. Ogasawara, K. Yamadate, M. Kobayashi, H. Endo, T. Fukuda, K. Yoshida, K. Terasaki, T. Inoue, and A. Ogawa, Effects of the free radical scavenger, edaravone, on the development of postoperative cognitive impairment in patients undergoing carotid endarterectomy, *Surg. Neurol., 64* (2005) 309-313.
[131] K. Toyoda, K. Fujii, M. Kamouchi, H. Nakane, S. Arihiro, Y. Okada, S. Ibayashi, and M. Iida, Free radical scavenger, edaravone, in stroke with internal carotid artery occlusion, *J. Neurol. Sci., 221* (2004) 11-17.
[132] M. Shibata, H. Hattori, T. Sasaki, J. Gotoh, J. Hamada, and Y. Fukuuchi, Activation of caspase-12 by endoplasmic reticulum stress induced by transient middle cerebral artery occlusion in mice, *Neuroscience, 118* (2003) 491-499.
[133] X. Qi, T. Hosoi, Y. Okuma, M. Kaneko, and Y. Nomura, Sodium 4-phenylbutyrate protects against cerebral ischemic injury, *Mol. Pharmacol., 66* (2004) 899-908.
[134] A. H. Lee, N. N. Iwakoshi, and L. H. Glimcher, XBP-1 regulates a subset of endoplasmic reticulum resident chaperone genes in the unfolded protein response, *Mol. Cell Biol., 23* (2003) 7448-7459.
[135] C. Kakiuchi, K. Iwamoto, M. Ishiwata, M. Bundo, T. Kasahara, I. Kusumi, T. Tsujita, Y. Okazaki, S. Nanko, H. Kunugi, T. Sasaki, and T. Kato, Impaired feedback regulation of XBP1 as a genetic risk factor for bipolar disorder, *Nat. Genet., 35* (2003) 171-175.
[136] W. Chen, S. Duan, J. Zhou, Y. Sun, Y. Zheng, N. Gu, G. Feng, and L. He, A case-control study provides evidence of association for a functional polymorphism -197C/G in XBP1 to schizophrenia and suggests a sex-dependent effect, *Biochem. Biophys. Res. Commun., 319* (2004) 866-870.
[137] I. Kusumi, T. Masui, C. Kakiuchi, K. Suzuki, T. Akimoto, R. Hashimoto, H. Kunugi, T. Kato, and T. Koyama, Relationship between XBP1 genotype and personality traits assessed by TCI and NEO-FFI, *Neurosci. Lett., 391* (2005) 7-10.
[138] C. A. Hetz and C. Soto, Stressing out the ER: a role of the unfolded protein response in prion-related disorders, *Curr. Mol. Med.,* 6 (2006) 37-43.

[139] L. Zhao, C. Longo-Guess, B. S. Harris, J. W. Lee, and S. L. Ackerman, Protein accumulation and neurodegeneration in the woozy mutant mouse is caused by disruption of SIL1, a cochaperone of BiP, *Nat. Genet., 37* (2005) 974-979.

In: Protein Misfolding
Editors: C. B. O'Doherty, A.C. Byrne

ISBN: 978-1-60456-881-3

Chapter IX

On the Nature and Possible Functions of Silent Modules in Proteins

Zakharia M. Frenkel **[*]** ***and Edward N. Trifonov***

Genome Diversity Center, Institute of Evolution, University of Haifa, Haifa 31905, Israel.

Abstract

We have recently discovered that 20 amino acid long sequence fragments of natural proteins, pair-wise similar to one another, can be organized in "walks" and networks, such that members of the same network may have rather different sequences, yet are related either functionally or structurally, or both. Traditional sequence comparisons would not classify such dissimilar sequence fragments, taken separately, as belonging to the same family. The sequence relatedness via many intermediates, i.e., via connections within the network in the sequence space, is the next, more advanced step in the development of sequence comparison methods. The networks suggest the kinship even between seemingly unrelated sequences, often belonging to different protein families.

The use of short fragments for the sequence comparisons reveals that some segments of a given protein sequence belong to large networks of the sequence space, i.e., have many close and distant relatives across species. At the same time, other segments are rather poorly represented in the space. As a result, the most informative sequence segments of a protein are identified, revealing a modular structure of the proteins.

Remarkably, such mapping by networks reveals existence of "silent" regions in the protein sequences. These regions belong to a very small networks, often of only one point size. These singular sequence fragments would be antonyms of conserved sequences, which appear in many proteomes. The first report of investigation of the nature of such silent regions in the protein sequences will be described in the chapter.

* Correspondence concerning this article should be addressed to: Dr. Z. M. Frenkel, Tel: (972-4)-8288097, (972-547)-959406; E-mail: zakharf@research.haifa.ac.il.

INTRODUCTION

Eight years ago Berezovsky et al. discovered that globular proteins are generally built from nearly standard size modules – closed loops of 25-30 residues [1]. That was confirmed by an independent study [2]. Actually, the term "closed loops" (vs "loops") had been introduced in 1995 by Ittah and Haas [3], though without any reference to their typical sizes. In protein structure literature the term "loop" is a misnomer, since it is normally understood as a connector between structural elements of protein (alpha-helices and beta-sheets), and it could be just a straight section of the chain. The term "closed loop", on the other hand, means a loop proper, with its ends closed.

Series of articles followed the original discovery of Berezovsky et al. [4-11]. It was found that the ends of the closed loops are largely hydrophobic [4,8], making tight contacts – locks [7,12]. This imposes certain constraints on the protein sequences – clusters of hydrophobic residues, at a distance 25-30 residues one from another. In other words, the closed loop structure of proteins is not just reflection of physics of the polypeptide chains, but also respective sequence design. It has been demonstrated that every globular protein, every protein fold can be presented, indeed, as a linear array of the consecutively connected closed loops [1,6]. Searching for most frequent consensus motifs in prokaryotic proteins, of the size 25-30 residues, resulted in the discovery of presumably ancient sequence/structure modules, prototypes Aleph, Beth,...Zayin [6,10,11]. Additional such motifs have been discovered after screening the protein sequences for omnipresent motifs (present in all or in a majority of prokaryotic proteomes) [13,14]. A full repertoire of the prototype elements would be needed, thus, for comprehensive description of the structure of modern proteins, that are built of the closed loop modules.

Interestingly, five of the original seven prototypes are found in ABC-transporters, all residing in one molecule [11]. The detailed analysis of one such molecule - ATP binding subunit of histidine permease of S. typhimurium – revealed that in addition to the prototype modules there are also closed loops in the molecule which do not have sequence analogs even amongst the large family of the ABC transporters. This was the first indication that many of the closed loops may serve rather as filling space elements. "Silent" regions in the protein sequences, that do not contain any frequent motifs are also detected by mapping the motifs along the sequences [15].

The studies on the modular structure of proteins gained substantially after the formatted protein sequence space has been introduced [16,17]. We have recently discovered that 20 amino acid long sequence fragments of natural proteins, pair-wise similar to one another, can be organized in "walks" and networks, such that elements of the same network may have rather different sequences, yet they are related either functionally or structurally [16-20]. Traditional sequence comparisons would not classify such dissimilar sequence fragments, taken separately, as belonging to the same family. The sequence relatedness via many intermediates, i.e., via connections within the network in the sequence space, is the next, more advanced step in the development of protein comparison methods.

The discovery of the formatted sequence space and of its non-trivial, revealing properties is both a breakthrough in protein research, opening a plethora of new directions and a challenge. As every principally new instrument it requires substantial efforts in "domestication" of this new tool. The sequence space topology, its relation to evolution, problem of completeness of the space – these are those issues that need an immediate attention.

One important issue is statistical evaluation of the classification identity between dissimilar fragments of the same network. Some of the fragments of the sequence space may connect to one of the networks purely accidentally, like, e. g., fragments with sequence errors, which do not really belong to the given network. The evaluation of the probability of accidental inclusion of the unrelated fragment to a network is, definitely, a problem waiting for solution. A whole new statistical tools have to be used for this brand new area – formatted protein sequence space. One estimate, perhaps, is worth mentioning. The total natural sequence space of, say 10^7 eukaryotic proteomes (for the sake of the estimate), of about 10^7 fragments each, makes 10^{14} fragments. On the other hand, full volume of all possible sequence fragments of this length is $20^{20} \approx 10^{26}$. That means that the natural sequence space occupies only about 10^{-12} of the total sequence space. This gives the probability for an arbitrary sequence to get into the natural space. The claim of identity (relatedness) of a given fragment to a given network becomes save when the fragment is connected to the network by more than one bond. That corresponds to majority of cases in the space of the walks and networks.

In principle, any size of the sequence fragments can be taken for construction of the sequence space. The size of the typical module, 25-30 residues, would be most natural choice. The sequences longer than that may contain two modules, not necessarily in the same combination. In this case, the sequence match would be compromised. Indeed, if the sequence A_1B_1, A_2B_2 and A_3C are compared (here A, B and C are different prototype modules), then the third one would match to the 1st and 2nd sequences only partially, to modules A, and, thus, may not get in the same network. The size 20 is well within the typical module size, and it also would fully cover smaller modules.

The use of the formatted sequence space with the close to module-size fragments reveals [19] that some segments of a given protein sequence belong to large networks of the sequence space, i.e., have many close and distant relatives across species. At the same time, other segments are rather poorly represented in the space, sometimes just by only one or few related fragments. As a result, the most informative (populated in space) sequence segments of a protein are identified, those that correspondent to the modules of the proteins.

The existence of the "silent" regions in the protein sequences requires an explanation. These often singular sequence fragments would be antonyms of conserved sequences, which appear in many proteomes. We proposed previously [19], that a possible role of the silent sequences may be to fill the sequence and 3D space between properly distanced and oriented functional modules. The silent sequences may have been originated from inserts of "random" sequences or from segments that have lost their original functionality.

The first report of investigation of the nature of such silent regions in the protein sequences will be described in the chapter.

RESULTS

1. Formatted Protein Sequence Space

We conducted all-to-all comparisons between the fragments (totaling over $9{\cdot}10^7$ fragments) and constructed networks of sequence relatedness at various thresholds of sequence identity, from 60% to 100% as described in methods. The limited studies on networks are reported in [18]. At a lower sequence similarity thresholds (60-65%) for connections of the points in the sequence space upto 10% of all the fragments assemble into one very large network around the sequences from the most common protein fold: P-loop NTPases [21,22]. At higher similarity thresholds this and other large networks split into a multitude of smaller clusters, whereas at even higher thresholds they split further into yet smaller ones, consisting of the fragments from the same functional and/or structural groups but still having rather different sequences. The splitting reflects more specific functional specialization of the proteins.

Several "silent modules" from two well studied protein types (ABC-transporters (1b0u, 1g29 and 1q12), and TIM-barrels (1tre and 2btm)) were selected for the analysis.

2. ABC Transporter 1b0u

The first silent region of 1b0u (at positions about 60-90) is located between the first and the second peaks (Figure 1A, Table 1), corresponding to conserved prototypes Aleph and Dalet (Figure 1B, red and green, respectively), described earlier [6,11]. As it is shown in the Figure 2A, the distance between these peaks, wall to wall, is rather variable keeping, however, a low level conservation of this region (data not shown). This holds for all proteins that possess the Aleph and Dalet type modules. The "silent module" sequences represented in the Figure 2A correspond to various ABC transporters from different Eubacteria.

Table 1. Silent regions of histidine permease

positions in proteins			Color	name of module
1b0u A	1g29 1	1q12 A		
28-60	25-58	25-57	Red	Aleph
61-92	59-79	58-72	Brown	
93-121	79-108	73-102	Green	Dalet
122-151	109-133	103-127	Blue	
152-190	134-174	128-168	Orange	Beth

The first silent region shows also differences in 3D-structure (Figure 3 A-C). The sequences corresponding to these structures are shown in Figure 4). Interestingly, although the structures of the inserts are different, the relative 3D positioning of the neighboring Aleph and Dalet modules is the same in all three cases (Figure 3 A-C).

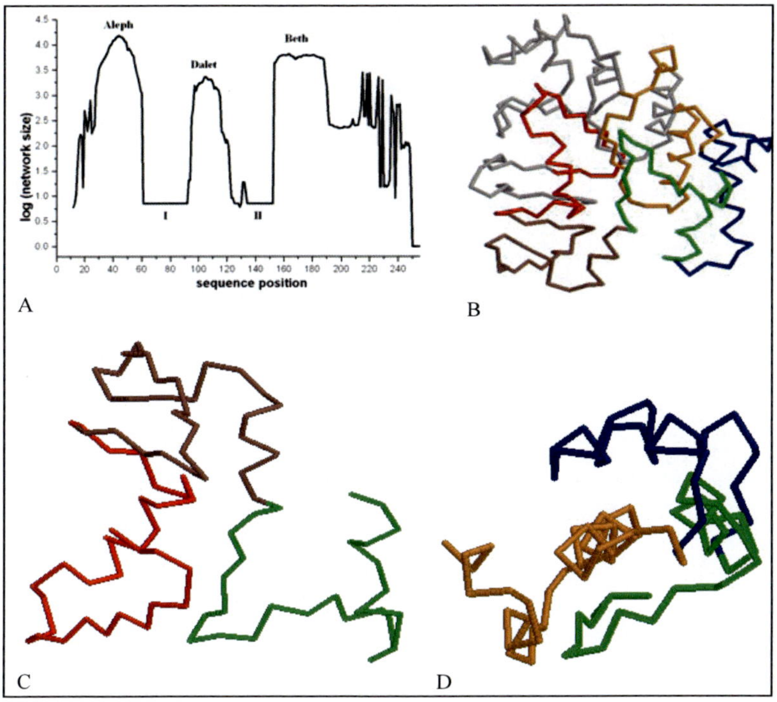

Figure 1. Silent regions of histidine permease. A) Sequence space (modular) presentation of ATP-binding subunit of histidine permease of S. typhimurium (1b0u, SwissProt code Q8Z511). For every point of the sequence the size of the cluster in the sequence space is indicated to which the respective sequence fragment (centered at the point) belongs. The networks with connecting threshold of 70% value are used. The peaks, corresponding to major modules Aleph, Beth and Dalet are labeled, respectively. The silent regions I and II are indicated. B) Backbone of the whole molecule of histidine permease (1b0u) with colored modules (Aleph (red), Beth (orange) and Dalet (green)) and correspondent silent regions (brown I and blue II) as described in Table 1. C) Backbone of the first silent region (I, brown) and neighboring modules Aleph (red) and Dalet (green). D) Backbone of the second silent region (II, blue) and neighboring modules Dalet (green) and Beth (orange).

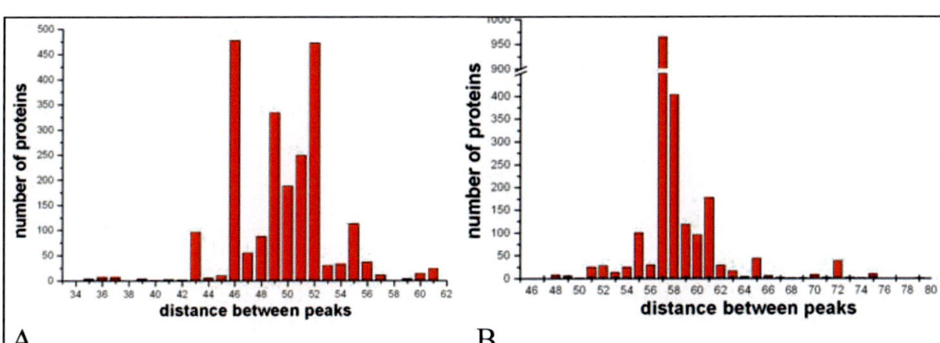

Figure 2. Size distribution of ABC-transporter silent regions. A) The silent region I. B) The silent region II. The histograms show the distribution of wall-to-wall distances between conserved modules Aleph and Dalet (A) and Dalet and Beth (B).

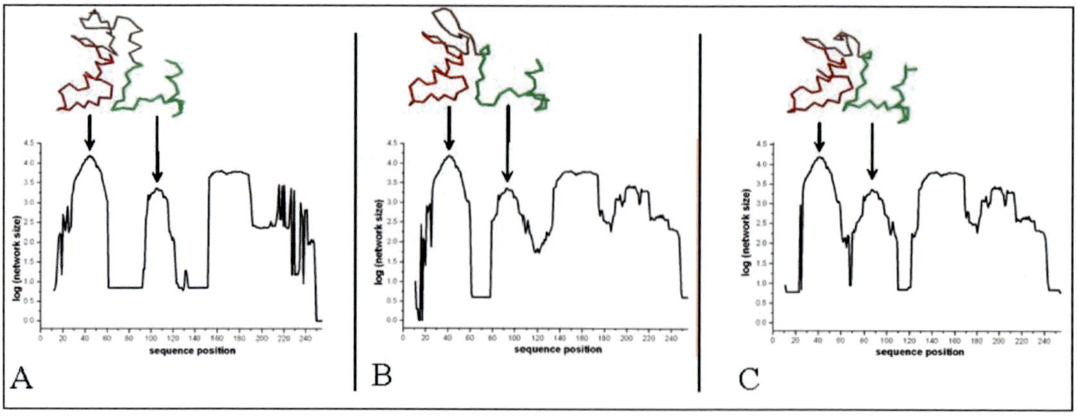

Figure 3. Comparison of the structures of the first silent regions of different size (see also Table I), in different ABC-transporters. A) 1b0u A; B) 1g29 1; C) 1q12 A; Note the conservation of relative positions of the modules Aleph and Dalet in 3D-structure.

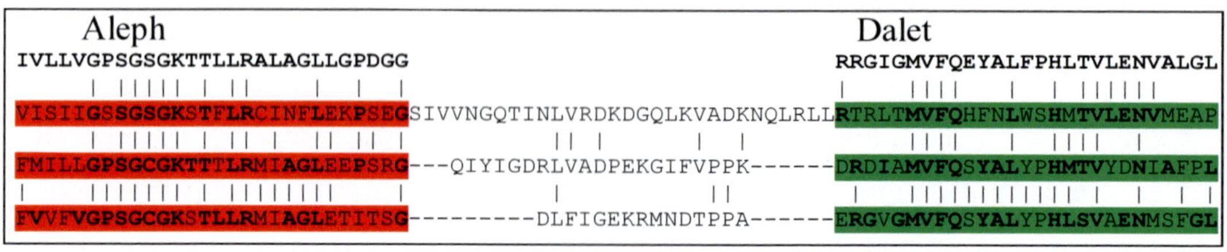

Figure 4. Comparison of sequences of the first silent regions of different sizes (see also Table I). On the top the "standard" prototype sequences are presented. The silent (non-shaded) regions are aligned to one another to optimize the match. The gaps are introduced, respectively.

The second silent region of 1b0u (region II, blue in Figure 1B, D) is located between the second and the third peaks (Figure 1A, Table 1), corresponding to described earlier [6,11] conserved prototypes Dalet and Beth (Figure 1B, green and orange, respectively). The size dispersion of this region is lower, than in the previous case (compare Figure 2B and Figure 2A). Similarly to the region I this silent region is also highly variable both sequence- and structure-wise (data not shown). In [11] this segment is described as Prototype VI, the only one of the seven original prototypes, that is not a closed loop. Considering the variations of that "module" both size-wise and sequence-wise, the original classification of this element as Prototype VI is, probably, erroneous. It appears rather as "silent" module

3. Tim-Barrel Proteins

As it turned out, the silent, low conservation regions of TIM-barrel proteins correspond to external alpha-helices. The typical examples (1tre from Escherichia Coli and 2btm from Bacillus cereus) are described in Figures 5-7. The size difference for the two cases shown is only one amino acid residue (Figure 7). It is reflected only in the presence or absence of an amino acid residue in the turn after the alpha-helix (see Figures 5 and 6, red). Again, the flanking structures appear "unaware" of the difference, being, essentially, identical (Figure 5C, D) and identically placed in 3D space.

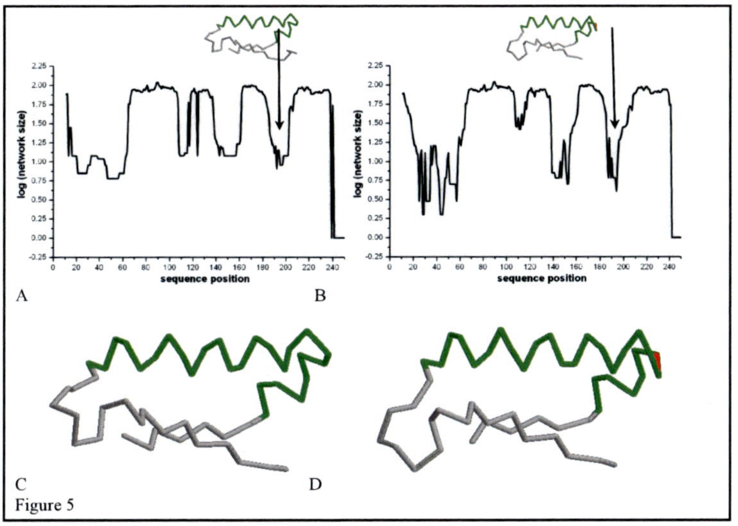

Figure 5. Silent regions of TIM-barrel proteins. A) Sequence space presentation of modules of Triosephosphate isomerase TIM from Escherichia Coli (1tre). B) Sequence space presentation of Triosephosphate isomerase TIM from Bacillus cereus (2btm). The last of four major silent regions in both cases are indicated by arrows, with respective structures above. C-D) 3D-structure comparison of the last silent regions in two TIM-barrel proteins. Note, the difference of one residue (in red) between C and D.

```
P60180 (2btm)
160               178                        205
ATVIAYEPIWAIGTGKSSSSADANEVCAHIRKVVAEAVSPEAAEAVRIQYGGSVKPENIKEYMAQSDIDGALVGGASLEPASFLGL
  |||||| ||||||||  | |  |   ||   |       || | ||||||   |  |  || ||||||||||||    |
GAVIAYEPVWAIGTGKSATPAQAQAVHKFIRDHIAKVDA-NIAEQVIIQYGGSVNASNAAELFAQPDIDGALVGGASLKADAFAVI
163               181                        207
```

Figure 6. Comparison of sequences of the last silent region of TIM-barrel proteins 2btm and 1tre, in green.

Discussion

Normally, when two protein sequences are compared, they are aligned to achieve maximal similarity, and the gaps are introduced at various positions in the sequences, to make the best possible match. All parts of the protein sequences are treated equally. The discovery and clear demonstration of modular structure of the protein sequences [11,19] suggest that this strategy has to be changed. In particular, the silent regions with no sequence conservation and with high structural variability, should not be part of the alignment. The conserved regions, the modules, have to be detected first, and aligned separately, module to module, to reveal relatedness of the compared proteins. The low conservation variable regions should be attended last, when the comparisons will come to details. Most significant information is confined, naturally, to the conserved (functional) regions, sequence modules. As Figures 4 and 6 illustrate, the alignment by conserved modules eliminates the problem of gaps, typical for traditional whole-sequence alignment. The gaps are naturally found in least conserved (if at all) "silent" regions.

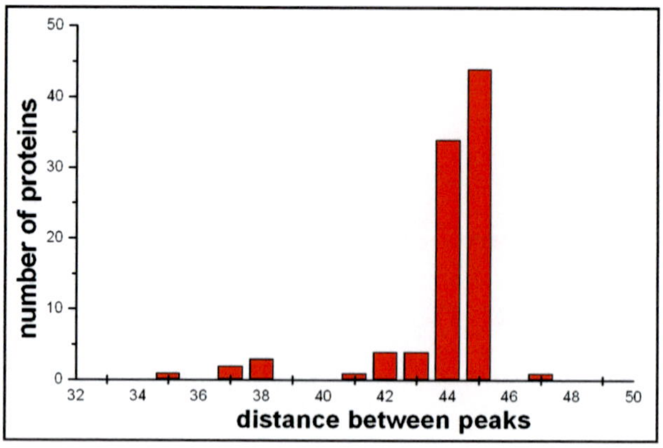

Figure 7. Size distribution of the last silent regions in various TIM-barrel proteins.

The “silent modules” or non-modules at all, are seemingly random sequence-wise. They may have originated either as inserts of “random” sequences, or evolved (randomized) from originally functional modules that have lost their functionality. The structure of the silent regions, on the other hand, is not random, though highly variable. From the examples described above it appears that the structures are under constraint to keep constant mutual orientation and distance between neighboring modules. We, thus, are confronted with the sequences, carrying only structural message – geometry of exiting and entering bonds of these alternative structures have to be identical, whatever is in-between. This feature of the silent regions, (and, probably, only this feature) apparently, is under selection pressure.

As truly random combination of the amino acid residues, probably, would not be able to convey this structural message, it remains to be found how it is encoded in the sequences, that is – what is this pure structural sequence code in the “silent” segments.

One can not exclude the possibility that the “silent” sequences are, actually, highly individual signatures of the proteins harboring them. The signatures could be crucial in very specific recognition events. They also may mean the “address” of the site they have to be delivered to. Finally, the sequences may correspond to very local environment of that given protein, such that certain interactions with its neighbors would be secured (or prevented). The generally outward positioning of the “silent” regions in the proteins suggests all the above interactions as quite likely explanation of the possible role(s) of the “silent” regions. One could think of an obvious analogy with non-coding regions in DNA. In the minds of molecular biologists the non-coding regions during last decades evolved from “junk” to respectful depot of transposons and highly conserved (and, thus, functional) elements. The “silence” of some protein fragments may well be a muted scream of some important functions, not yet in reach of our understanding.

METHODS

A database that contains proteomic positions of all connected sequence fragments of this space was developed by the following algorithm.

Note, that if two given sequences of size N aa are identical to one another at a threshold T (in %), at least K amino acids ($1 \leq K \leq 0.01 \times N \times T$) should be identical with a specific configuration C. For example, for N = 10 and T = 90% (i.e., 10 aa with only one permitted mismatch) for K = 3, the possible configurations C are {XXX}, {X-XX} or {XX-X} ('X' match, '-' mismatch). In many cases (for some N, K, and T values), when two sequences are compared certain configurations are always present. For example, for N = 10 and K = 3, configuration C = {XXX} is present for all 70% < T ≤ 100%. Generally, for every combination of N, K, and T values a set of Nc configurations exists, such that for every pair of compared sequences at least one of the Nc configurations is present. For example, in a case of {N = 10, K = 3, T = 70%} Nc = 2. The set Nc consists of two configurations: C1 = {XXX} and C2 = {XX-X}.

Essentially, the algorithm helps to avoid all-to-all comparisons. The catalogues are built from all oligopeptides (continuous or interrupted) corresponding to various configurations, as above. Such catalogues, together with proteomic locations of the peptides are of the size suitable for operation memory of PC. Instead of all sequences being compared, only those sequences that share one or more oligopeptides of the database, and, thus listed in the same catalogue(s), are compared.

For example, for N = 10 and T > 70% there is only one universal configuration C = {XXX}. Taking the first tripeptide of a given fragment, one would find only small proportion of other fragments with the same tripeptide (in the same position), of the order of 20^{-3}. With a total of 8 triplets at hand, the gain in the volume of comparisons is about 1000-fold.

For this work, with N = 20, T = 60%, tetramers (K = 4) were used. Considered configurations are C1 = {XXXX}, C2 = {XXX-X}, C3 = {XX-XX}, C4 = {XXX--X}, C5 = {XX-X-X}. At least one of these configurations should be present when any two sequences (N = 20, T = 60%) are compared. This full set of all configurations is generated by exhaustive combinatorics.

The initial database was used for the generation of connected networks of the fragments, such that every fragment of the network is similar (by simple match) to one or more of the other fragments of the network, at the given identity threshold (60%). The initial database has a very large size (more than 5 Gb) and could not be placed in the operative memory of the personal computer. Therefore, at the first reading of this database, a list of trees and a list of nodes were created. Each record on the list of trees contained its order number as well as the order number of another connected tree, located upstream in the list (if any). Each record on the list of nodes contained sequence number and position of corresponding sequence fragments and number of trees to which this node belongs.

Every new node in the reading process, if not assigned to any tree earlier, was assigned to a new tree and added to the list. Every node directly connected to the first one was assigned as belonging to a respective tree. If one of these connected nodes has been already assigned to a different tree, these two trees were connected.

Other stages of the network generation included combining trees and renumbering. The final lists of trees and nodes are much smaller than the initial database of connections and can be placed in the operative memory of computer. The whole procedure takes a relatively short calculation time (about 1 h). The sequences with simple repeats (more than 6 occurrences of the same amino acid in the fragment) and sequences, many times repeated in the same protein, were excluded from consideration.

The database generated for these 20 amino acid fragments and their immediate connections (at the 60% identity threshold), specifying their locations in the protein sequences, are available at http://genome.haifa.ac.il/~zfrenkel/DATABASES1/.

Acknowledgement

This work was supported by the Center for Complexity Science grant GR2006-018.

References

[1] Berezovsky, I. N., Grosberg, A. Y. & Trifonov, E. N. (2000). Closed loops of nearly standard size: common basic element of protein structure. *Febs Letters 466*, 283-286.

[2] Lamarine, M., Mornon, J. P., Berezovsky, I. N. & Chomilier, J. (2001). Distribution of tightened end fragments of globular proteins statistically matches that of topohydrophobic positions: towards an efficient punctuation of protein folding? *Cellular and Molecular Life Sciences 58*, 492-498.

[3] Ittah, V. & Haas, E. (1995). Nonlocal Interactions Stabilize Long-Range Loops in the Initial Folding Intermediates of Reduced Bovine Pancreatic Trypsin-Inhibitor. *Biochemistry 34*, 4493-4506.

[4] Trifonov, E. N., Kirzhner, A., Kirzhner, V. M. & Berezovsky, I. N. (2001). Distinct stages of protein evolution as suggested by protein sequence analysis. *Journal of Molecular Evolution 53*, 394-401.

[5] Berezovsky, I. N. & Trifonov, E. N. (2002). Loop fold structure of proteins: Resolution of Levinthal's paradox. *J. Biomol. Struct. Dyn. 20*, 5-6.

[6] Berezovsky, I. N. & Trifonov, E. N. (2002). Flowering buds of globular proteins: transpiring simplicity of protein organization. *Comparative and Functional Genomics 3*, 525-534.

[7] Berezovsky, I. N. & Trifonov, E. N. (2001). Van der Waals locks: Loop-n-lock structure of globular proteins. *Journal of Molecular Biology 307*, 1419-1426.

[8] Berezovsky, I. N., Kirzhner, V. M., Kirzhner, A. & Trifonov, E. N. (2001). Protein folding: Looping from hydrophobic nuclei. *Proteins-Structure Function and Genetics 45*, 346-350.

[9] Trifonov, E. N. & Berezovsky, I. N. (2003). Evolutionary aspects of protein structure and folding. *Current Opinion in Structural Biology 13*, 110-114.

[10] Berezovsky, I. N., Kirzhner, A., Kirzhner, V. M., Rosenfeld, V. R. & Trifonov, E. N. (2003). Protein sequences yield a proteomic code. *J. Biomol. Struct. Dyn. 21*, 317-325.

[11] Berezovsky, I. N., Kirzhner, A., Kirzhner, V. M. & Trifonov, E. N. (2003). Spelling protein structure. *J. Biomol. Struct. Dyn. 21*, 327-339.

[12] Aharonovsky, E. & Trifonov, E. N. (2005). Sequence structure of van der Waals locks in proteins. *J. Biomol. Struct. Dyn. 22*, 545-553.

[13] Sobolevsky, Y. & Trifonov, E. N. (2005). Conserved sequences of prokaryotic proteomes and their compositional age. *Journal of Molecular Evolution 61*, 591-596.

[14] Sobolevsky, Y. & Trifonov, E. N. (2006). Protein modules conserved since LUCA. *Journal of Molecular Evolution 63*, 622-634.

[15] Aharonovsky, E. & Trifonov, E. N. (2005). Protein sequence modules. *J. Biomol. Struct. Dyn. 23*, 237-242.

[16] Frenkel, Z. M. & Trifonov, E. N. (2007). Walking through protein sequence space. *Journal of Theoretical Biology 244*, 77-80.

[17] Frenkel, Z. M. & Trifonov, E. N. (2007). Walking through the protein sequence space: Towards new generation of the homology modeling. *Proteins-Structure Function and Bioinformatics 67*, 271-284.

[18] Frenkel, Z. M. & Trifonov, E. N. (2007). Evolutionary networks in the formatted protein sequence space. *Journal of Computational Biology 14*, 1044-1057.

[19] Frenkel, Z. M. & Trifonov, E. N. (2008). From protein sequence space to elementary protein modules. *Gene 408*, 64-71.

[20] Sobolevsky, Y., Frenkel, Z. M. & Trifonov, E. N. (2007). Combinations of ancestral modules in proteins. *Journal of Molecular Evolution 65*, 640-650.

[21] Wolf, Y. I., Brenner, S. E., Bash, P. A. & Koonin, E. V. (1999). Distribution of protein folds in the three superkingdoms of life. *Genome Research 9*, 17-26.

[22] Hegyi, H., Lin, J., Greenbaum, D. & Gerstein, M. (2002). Structural genomics analysis: Characteristics of atypical, common, and horizontally transferred folds. *Proteins-Structure Function and Genetics 47*, 126-141.

In: Protein Misfolding
Editors: C. B. O'Doherty, A.C. Byrne
ISBN: 978-1-60456-881-3

Short Communication A

New Perspectives in Small-Angle Scattering to Study Unstructured States of Proteins

Pau Bernadó[1] and Dmitri I. Svergun[2,3]
[1]Institute for Research in Biomedicine, Parc Científic de Barcelona
Baldiri Reixac 10-12, 08028 Barcelona, Spain
[2]European Molecular Biology Laboratory, Hamburg Outstation
Notkestrasse 85, 22603 Hamburg, Germany;
[3]Institute of Crystallography, Russian Academy of Sciences
Leninsky pr. 59, 117333 Moscow, Russia

Abstract

Small-angle scattering (SAS) of X-rays (SAXS) and neutrons (SANS), have been widely used to study the overall structure of folded and unfolded proteins in solution. SAS measurements of unstructured states are normally analysed in terms of radii of gyration (R_g) that are either monitored during protein folding/unfolding processes or compared with those predicted for native states or random coil models. Our groups have recently developed a new methodology, the so-called Ensemble Optimization Method (EOM), analysing the scattering data in terms of ensembles of coexisting conformations, which, when treated collectively fit the experimental profile. This strategy allows the description of size properties of disordered proteins in solution as distributions within their ensembles. Therefore, the traditional single value averaged R_g description of a flexible state can be converted into a more detaied picture that is the density of accessible R_g values. Thanks to this, even short fragments containing secondary structure, or presence of transient interactions between remote regions of the chain can sometimes be identified. The localization of these structural features can be addressed through the simultaneous analysis of profiles measured for different fragments (e.g. deletion mutants) of the protein.

The EOM strategy offers more detailed structural information of unstructured states and is highly complementary with other structural biology techniques such as Nuclear Magnetic Resonance (NMR), which provide access to the conformational sampling at residue level, or circular dichroism, sensitive to the presence of secondary structural elements. Applied alone or in combination with other structural information, the EOM can provide new insights into the study of the protein folding pathways or the conformational changes upon environmental changes, and we envision it as the method of choice for the analysis of SAS data of highly flexible biomolecules.

Introduction

The unstructured state of proteins has intrigued many researchers for decades. The knowledge of the rules governing the unfolded state ensures a better comprehension of the protein folding pathway, the thermodynamic stability of proteins and the process of amyloid fibrils formation [1,2]. Therefore, the unfolded state has been mainly studied for globular, well folded proteins under harsh conditions such as extreme temperatures or "good" solvents such as urea or guanidinium chloride. This interest has been recently renewed with the discovery of a new family of proteins, the so-called Intrinsically Unfolded Proteins (IUPs) [3,4]. These proteins are fully active and play important biological roles in several metabolic and signal transduction pathways even though they lack a permanent tertiary structure [5,6,7].

A complete structural characterization of unfolded or partially unfolded proteins must account for the behaviour of the conformations at residue level (local properties), and features related to the protein size and shape (global properties). Local properties tell us about the formation and population of nascent secondary structural elements such as α-helices, β-strands, or turns, and more generally about the conformational nature of the random-coil. Global properties reflect the volume occupied by the unstructured protein and, therefore, monitor its degrees of compactness and unfolding in solution. It has been shown that both local and global structural properties are crucial to understand the sequential folding pathway of globular proteins [8,9]. In addition, partially folded regions in IUPs are involved in biologically relevant protein-protein interactions, and their population finely modulates the thermodynamic and kinetic properties of these interactions [10,11]. It has also been shown that long-range interactions in IUPs can prevent pathological aggregation [12,13]. Although different in nature, local and global features are intrinsically coupled, for instance the presence of a certain population of a secondary structural elements will be reflected in the overall size and shape of the protein. Therefore, for the study of unstructured proteins, biophysical methods or an appropriate combination of them able to simultaneously describe both sources of structural information become mandatory. The main problem for the interpretation of biophysical data is the inherent flexibility of unstructured states, such that any feasible model requires a representation of the protein as an ensemble of accessible conformations. Therefore, experimentally measured data must be interpreted as an average of the property over all the members of the ensemble. This scenario is opposed to the single configuration modelling traditionally used for the characterization of globular proteins.

Small-angle scattering (SAS) of X-rays (SAXS) and neutrons (SANS), is a well-established method for the structural characterization of biomolecules in solution [14,15].

Recent developments in high brilliance synchrotron beamlines, and in software for a rational interpretation of SAS profiles in terms of three-dimensional models have lead to a recent renaissance of this method in structural biology [16,17]. SAS has also traditionally been the method of choice to characterize the global properties related to size and shape, and degree of compactness of unstructured states in solution [18]. Probably the most common strategy to identify an unfolded protein by SAS is a so-called Kratky representation of the scattering profile that also provides a qualitative indication of the degree of compactness of the protein. As displayed in figure 1, in a Kratky representation the scattering intensity I(s) multiplied by the square of the momentum transfer $s = 4\pi \sin\theta/\lambda$, where 2θ is the scattering angle and λ is the radiation wavelength. Globular and compact molecules present a pronounced peak in the Kratky's representation, whereas unstructured states show a completely different behaviour with a continuous increase of $s^2 \cdot I(s)$ with the angle.

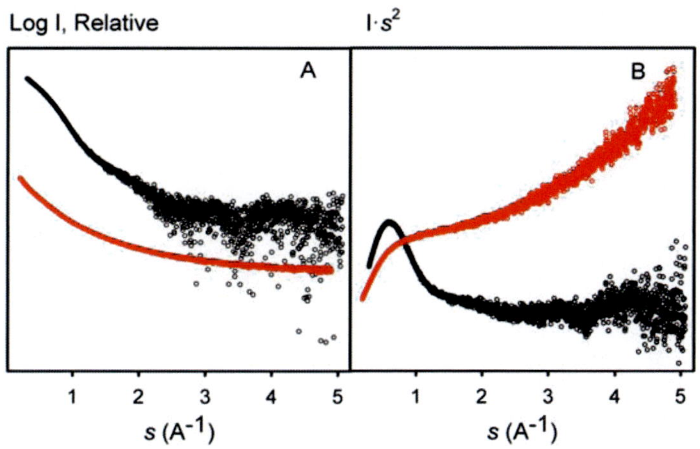

Figure 1. Scattering profiles (A) and Kratky representations (B) of a folded protein, Bovine Serum Albumin (black), and a denatured protein, lysozyme in 8M urea and 100mM DTT (red). Kratky representations are a useful tool to qualitatively distinguish between globular and unstructured proteins in solution.

As a quantitative characteristic, the radius of gyration of the protein (R_g), can be derived with the Guinier approximation assuming that at very small angles ($s < 1.3/R_g$) the intensity is represented as $I(s) = I(0) \exp(-(sR_g)^2/3)$ [19]. For unstructured proteins, alternative Debye approximation [20] is also used, which may provide more accurate R_g estimates. The experimental R_g is a single value representation of the size of the molecule yielding an average (so-called z-average, [21]) over all accessible conformations in solution. Very often, the quantitative analysis for unstructured proteins is limited to the monitoring of changes in R_g, which, of course, dramatically reduces the structural information extracted from SAS data to a single parameter. A classical quantitative R_g interpretation in terms of the Flory's equation [22] relates it to the length of the protein chain through a power law

$$R_g = R_0 \cdot N^{\nu} \quad (1)$$

where N is the number of residues in the polymer chain, a constant R_0 depends on several factors, in particular, on the persistence length, and ν is an exponential scaling factor. For an excluded-volume polymer Flory estimated ν to be approximately 3/5, and more accurate estimates established a value of 0.588 [23]. A recent compilation of R_g values measured for 26 chemically denatured proteins sampling broad range of chain lengths found a ν value of 0.598 ± 0.028 [24], in good agreement with hydrodynamic data [25], and a R_0 value of 1.927 ± 0.27. Several groups have addressed the computation of the averaged R_g values from ensembles representing the unfolded states of different complexity and the theoretical predictions yielded good agreement with the experimental data [26-29]. These results suggest the random coil nature of the chemically denatured proteins, at least in terms of the R_g parameter. In contrast, Nuclear Magnetic Resonance (NMR) detects certain amount of secondary structural elements or even a native-like topology in the denatured state [8,30-33]. These apparently apposite observations gave rise to the so-called "reconciliation problem" [24,26,34]. A theoretical study shows, by simulating unfolded protein ensembles, that small amounts of structure, potentially measurable by NMR, may not significantly alter the average R_g [35]. These simulations suggest that R_g is not sufficiently sensitive to small amounts of residual structure in unfolded proteins. To add further arguments to the discussion, a recent NMR study based on the measurement of several Residual Dipolar Couplings (RDC) along ubiquitin backbone suggested that the conformational sampling in chemically denatured proteins could be different from that found in IUPs [36]. As a consequence, even in the absence of transiently structured regions, Flory's law may not be appropriate parameterization of R_g for natively unstructured proteins. From all this data, it seems clear that the study of the unfolded proteins by SAS would benefit from approaches, where additional information is used including scattering profile regions beyond Guinier approximation. The inclusion of higher resolution information could in principle allow one to observe and describe partial structuring in the proteins.

The characterization of the scattering profiles of highly flexible proteins at molecular level has been hampered by the necessity of accoutning for multiple configurations as described above. Some studies compared experimental scattering profiles with simulated ones calculated as the average of medium to large ensembles of conformations generated under a certain model of unfolded state [37-40]. This trial-and-error approach allows one to validate theoretical models but can of course not be generally applied for model-free analysis.

The Ensemble Optimization Method (EOM)

Our groups have recently developed a new methodology, the so-called Ensemble Optimization Method (EOM), that allows one to overcome some of the above limitations [41]. Using this strategy the experimental SAS data are assumed to arise from an (undetermined *a priori*) number of coexisting conformational states. In this way the concept of ensemble is introduced into the analysis of SAS data generalizing the traditional 'single state' methods. The conformations belonging to the final ensemble are selected using a simple genetic algorithm (GA) based on the scattering patterns computed from a large representative random pool. In this sense, the final result is not a trial-and-error but rather a data-driven optimised ensemble strategy. Instead of a single averaged R_g value, this strategy provides a distribution of the R_g's in the selected ensemble, which is undoubtedly a step forward in the analysis of SAS data for flexible proteins. The price to pay in order to optimize the large number of degrees of freedom by EOM is that assumptions are required to generate the pool of unfolded chains. The randomization model is also used to establish a threshold for the later quantitative interpretation of the results in terms of individual selected structures, and in this sense the EOM is not entirely model-independent. It is worth mentioning that EOM provides a new source of structural information for unstructured and multidomain proteins [42,43]. Only single molecule Fluorescence Resonance Energy Transfer (FRET), able to measure site-to-site distance distributions, can provide similar structural constraints [44,45]. The major features of the EOM will be discussed in the following paragraphs.

The Principle of the EOM Algorithm

Figure 2 schematically displays the algorithm designed to perform the optimization. We represent the molecule in solution by an ensemble containing N different conformations (in practice, N typically ranges between 20 and 50). The scattering data from such an ensemble is computed by averaging the individual scattering patterns from the conformers. The appropriate ensemble is selected from a pool containing a large number $M>>N$ of random conformers which should cover the conformational space of the molecule. A Monte-Carlo based search is then employed to select subsets of configurations that after averaging their individual scattering profiles fit the experimental data. We can assume that the subsets are uniformly populated, so that the intensity of a subset I(s) containing N conformers is

$$I(s) = \frac{1}{N}\sum_{n=1}^{N} I_n(s) \tag{2}$$

Here, $I_n(s)$ is the scattering from the n-th conformer. To speed up the calculations, the scattering curves from all the structures in the pool are first pre-computed using the program CRYSOL [46] using default parameters and the subsequent selection operators are preformed using these patterns, and not the structures.

The subset selection is performed using a GA, where each subset (chromosome) is composed of N scattering profiles (genes) corresponding to different conformers. In the first generation C=50 chromosomes are generated by selecting conformations/curves randomly from the pool. In each generation G these C chromosomes are submitted to two genetic operators: *random mutation* and *crossing*. In *random mutation* the genes of each chromosome were exchanged for others either from the pool or from the chromosomes of the same generation. In the *crossing* operator two chromosomes were selected randomly and their genes were exchanged. Details about the parameters describing the two operators can be found in the original paper [41]. At the end of these two genetic operations the population is composed of $3C$ chromosomes. For each chromosome, the average (equation 2) of its individual SAS profiles is compared with the experimental scattering yielding the fitness function (equation 3).

$$\chi^2 = \frac{1}{K-1}\sum_{j=1}^{K}\left[\frac{\mu I(s_j) - I_{\exp}(s_j)}{\sigma(s_j)}\right]^2 \tag{3}$$

where $I_{exp}(s)$ is the experimental scattering, K is the number of experimental points, $\sigma(s)$ are standard deviations and μ is a scaling factor. The best C chromosomes yielding the lowest discrepancy with the experimental profile, are selected for further evolution, typically for up to G=1,000 to 5,000 generations. The chromosome with the best agreement to the experimental data is collected for further analysis. Normally the complete process is repeated R = 200 times in order to collect 10,000 final conformations.

The structural examination of the final ensemble is performed by analysing the distribution of low-resolution parameters, R_g, maximum distance (D_{max}) and anisotropy, of the selected conformations. The relevance of the selected conformations with respect to the behaviour of the protein in solution will be discussed in following paragraphs.

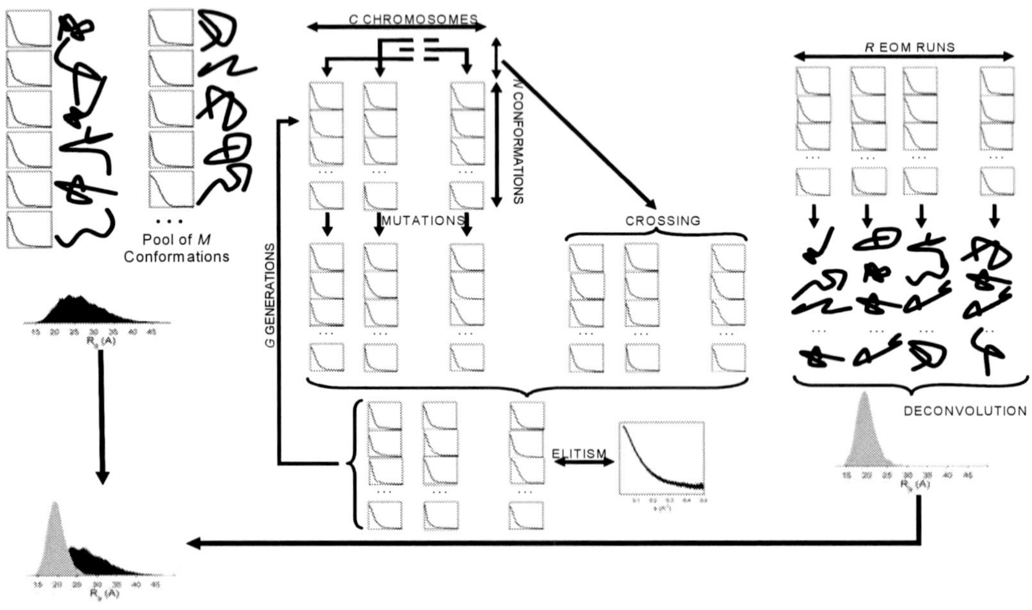

Figure 2. Schematic representation of the EOM algorithm and analysis of the results in terms of R_g distributions. The M conformations/curves of the pool, left part of the figure, are used to generate the initial C chromosomes and to feed the genetic operators along the GA process. The analysis of the resulting conformations is displayed on the right part of the scheme, and the R_g distribution is compared with that derived from the pool.

How Much Structural Information Is Coded in EOM Results?

An unstructured protein in solution samples an astronomical amount of conformations. It may seem that a subpopulation with $N\approx50$ is far too small to explain the conformational behaviour of a flexible protein in solution. However, the aim of the method is deriving the low resolution structural information of the protein in solution describing the size and shape, i.e. R_g, D_{max} and anisotropy, rather than conformational details at residue level. In this sense the pool of $M>>N$ potential conformations must represent a complete survey of global characteristics that a chain of a certain length can adopt in solution. The structures selected by the EOM are not claimed to be the most populated conformations in solution, they do not have even to exist in solution. These structures must be understood as a tool to describe the size and shape distributions sampled by the unfolded molecule in solution. In fact, selected ensembles derived from repeated EOM runs starting from different random seeds normally contain different conformations but they all provide similar R_g, D_{max} and anisotropy distributions. Therefore, the algorithm is able to find equivalent minima in terms of distributions combining different ensembles of conformations.

Although the distributions provided by EOM yield a major improvement over traditional approaches that condense all structural characteristics in the averaged R_g value, one should bear in mind that the EOM results are still limited in resolution. This limitation is imposed by the restricted information content of SAS technique reporting on the size and shape of particles but not on the higher resolution characteristics. An advantage of such restricted information content is the ability of physical models containing relatively small ensembles to adequately describe the entire system [41]. Other biophysical observables with higher information content, such as the RDC (probing the conformational sampling at atomic level), require much larger ensembles ($\approx 10^4$ conformers) to reach reasonable levels of convergence [28,39].

The EOM-derived distributions offer new opportunities for the structural characterization of unfolded proteins in solution, and modifications upon environmental changes can be now more easily assessed. However, interpretation in absolute terms still requires some sort of threshold distributions arising from a random coil model. Through this comparison, experimental *vs* random coil distributions, one can detect the presence of nascent secondary structural elements or low populations of transient long-range contacts as alterations of the distributions with respect to the threshold one. Local transient structures display broader distributions of R_g and normally shifted towards larger values. Long-range contacts and partial tertiary structure display more compact distributions than the random coil one.

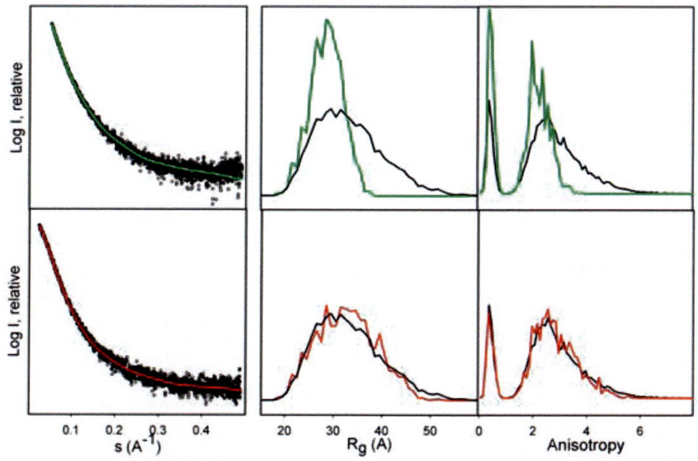

Figure 3. EOM fits to the partially reduced (green) and completely reduced (red) lysozyme unfolded in 8M urea solutions (left panel). Middle and right panels: the R_g and anisotropy distributions, respectively, from the pool (black) and optimized ensembles for both lysozyme samples. Figure taken with permission from P. Bernadó, E. Mylonas, M.V. Petoukhov, M. Blackledge, and D.I. Svergun, *J. Am. Chem. Soc.* 2007, 129, 5656-5664. Copyright 2007, American Chemical Society.

Figure 3 displays an example EOM analysis of two different unfolded states of lysozyme. The SAXS curve of the denatured lysozyme sample in 8M urea and moderate reducing conditions (10mM DTT) yields a R_g = 26.3 Å pointing to a highly unstructured system (R_g = 15.0 Å in the folded state). Interestingly the EOM analysis, figure 3 top panels, clearly indicates that the protein is not completely unfolded, and more compact conformations are still present in solutions when compared with distributions derived from the pool (black distributions). When the amount of the reducing agent was increased by 10 times (bottom panels), experimental distributions coincide with those calculated from the pool, suggesting that the protein is now fully unstructured. The significant difference observed between both samples when monitored with EOM is in contrast with a relatively moderate increase of R_g, (to 30.0 Å), with the addition of extra DTT. These results can be readily explained by partial reduction of the four lysozyme disulphide bridges in moderate reducing conditions [41].

The previous example highlights the importance of having a good description of the unfolded state to compare and for further interpretation of the EOM results. However, as explained in previous paragraphs, structural models for the unfolded (random coil) state are under debate. We have addressed this problem by using the flexible-Meccano program to generate the conformations of the unfolded state [39]. The conformational sampling used by this program has been derived from a coil library obtained from a large database of high-resolution X-ray structures [47] from which all residues in α-helices and β-sheets have been removed. This program is capable of describing the RDC measured in several IUPs, that probes conformational fluctuations along the backbone, and therefore seems to be an adequate model for the unfolded state of proteins [12,39,48-50].

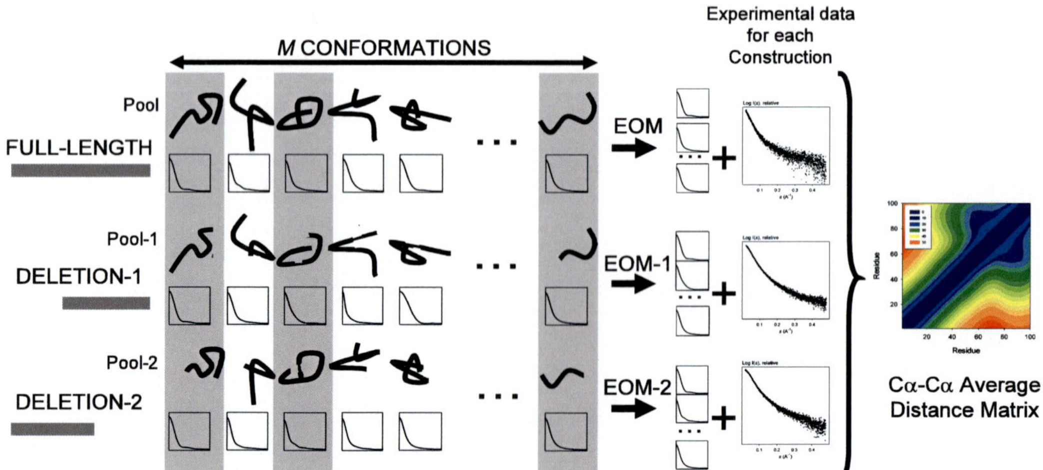

Figure 4. Schematic representation of the EOM strategy applied simultaneously to multiple deletion mutants. Grey boxes represent an ensemble of curves/scattering profiles selected along the GA. An example of the C_α-C_α average distance matrix that can potentially be obtained via this analysis is also displayed, which may permit one to identify long-range contacts along the chain.

Although EOM allows for detection of overall structuring phenomena the exact regions responsible for conformational restriction remains elusive. A strategy to partially overcome this limitation was designed and tested with synthetic data in the original paper [41]. The scattering profiles from different deletion mutants of the full-length protein (if available) can be fitted simultaneously with EOM, and the sum of the individual figures of merit drives the optimization, see figure 4. The aim of this approach is to distinguish between the contributions from different chain fragments to the SAS data and to identify the regions responsible for the increase or decrease in compactness of the protein. In other words the resolution of the technique is improved with the addition of multiple experimental curves. Note that this strategy is only valid if the structural elements of the full length protein remain intact in the deletion mutants.

Future Perspectives

The possibility of modelling SAS data in terms of ensembles of conformations opens new perspectives in the study of unstructured and partially folded proteins and further enhances the potential of SAS in the study of protein folding. The EOM, through the distribution of structural parameters such as the R_g, should allow a better detection of structured regions or long-range contacts along the protein chain. This can have important applications in the biophysical characterization of IUP, a growing class of biomolecules difficult to address with conventional structural techniques [51]. EOM is also promising in the study of the protein folding pathway and characterization of the folding intermediates that contain nascent secondary structural elements of transient long-range contacts previous to the folding collapse [52], and could eventually lead to misfolding and aggregation [53,54]. A number of kinetic folding studies have been reported [55-57], and EOM can significantly contribute to a structure based monitoring of the transformation along the pathway.

The nature of the unfolded state, the starting or the ending points in folding studies, and the physicochemical properties of a random coils can also be addressed with EOM. This could also shed some light to controversial aspects of the random coil definition such as the conformational explorations and the global properties of the ensemble.

The structural parameter distribution derived from EOM is an intuitive description of the global properties of a biomolecule providing a more direct insight into the structural features. These distributions can be potentially modelled with some *a priori* knowledge of the system, for instance, folding pathways can be validated or discarded if the high-resolution structure of the folded state is available and/or structural models of intermediate states can be calculated. Secondary structure predictions for IUPs can also be tested.

The structural information derived from SAS using EOM can be complemented by other structural techniques. Especially promising is the synergy with NMR traditionally applied to the study of unfolded and partially folded proteins [58-60]. The combination of the global information derived by SAS with NMR, that can probe local conformations through chemical shifts, scalar couplings, homonuclear Overhauser effect (nOe) and especially with RDC, can be an extremely powerful tool to derive more accurate structural models of highly flexible molecules. In addition NMR can detect long-range contact networks through Paramagnetic Relaxation Experiments (PRE) where specific residues are chemically modified to attach a paramagnetic probe inducing and enhancing relaxation to the proximal amino acids [61]. Although the ensemble-averaged nature of this biophysical data is equivalent to scattering profiles [62-64], PRE provides site-specific contact information highly complementary to global properties obtained with SAS. In addition, this residue-specific information can help in breaking the intrinsic symmetry problems (i.e. inability to distinguish between residues) of SAS data for unfolded systems. Other techniques such as circular dichroism (CD) and fluorescence energy transfer (FRET) can also provide complementary structural information. CD is sensitive to the presence of secondary structural elements, whereas FRET allows one to assess the distance between two chromophores covalently attached to the protein.

In summary we think that the EOM applied to SAS data provides a new tool for the biophysical characterization of unstructured and partially folded protein states, which are difficult to comprehensively describe using a single technique. The EOM provides a new complementary piece of structural information, the distribution of the accessible size properties in solution. This information can also be integrated with other biophysical data to derive more accurate and complete structural models for molecular systems that play pivotal roles in biological processes.

Acknowledgements

P.B. holds a Ramón y Cajal contract that is partially financed by the Spanish Ministry of Education and by funds provided to the Institute for Biomedical Research by the Generalitat de Catalunya. P.B. acknowledges funds from the Spanish Ministry of Education-FEDER (BIO2007-63458). D.S. acknowledges support by the European Commission through a Design Study SAXIER, grant number RIDS 011934.

References

[1] Shortle, D. The denatured state (the other half of the folding equation) and its role in protein satability. *FASEB J.* 1996, 10, 27-34.

[2] Dobson, CM. Experimental investigation of protein folding and misfolding. *Methods* 2004, 34, 4-14.

[3] Wright, PE; Dyson, HJ. Intrinsically unstructured proteins: Re-assessing the protein structure-function paradigm. *J. Mol. Biol.* 1999, 293, 321-331.

[4] Tompa, P. Intrinsically unstructured proteins. *Trends Biochem. Sci.* 2002, 27, 527-533.

[5] Dunker, AK; Brown, CJ; Lawson, JD; Iakoucheva, LM; Obradovic, Z. Intrinsic disorder and protein function. *Biochemistry* 2002, 41, 6573-6582.

[6] Dafforn, TR; Smith, CJI. Natively unfolded domains in endicytosis: Hooks, lines and linkers. *EMBO Rep.* 2004, 5, 1046-1052.

[7] Dyson, HJ; Wright, PE. Intrinsically unstructured proteins and their functions. *Nature Rev.* 2005, 6, 197-208.

[8] Klein-Seetharaman, J; Oikawa, M; Grimshaw, SB; Wirmer, J; Duchardt, E; Ueda, T; Imoto, T; Smith, LJ; Dobson, CM; Schwalbe, H. Long-range interactions within a nonnative protein. *Science* 2002, 295, 1719-1722.

[9] Vendruscolo, M; Paci, E; Karplus, M; Dobson, CM. Structures and relative free energies of partially folded states of proteins. *Proc. Natl. Acad Sci. USA* 2003, 100, 14817-14821.

[10] Sugase, K; Dyson, HJ; Wright, PE. Mechanism of coupled folding of an intrinsically disordered protein. *Nature*, 2007, 447, 1021-1025.

[11] Fuxreiter, M.; Simon, I.; Friedrich, P.; Tompa, P. Preformed structural elements feature in partner recognition by intrinsically unstructured proteins. *J. Mol. Biol.* 2004, 338, 1015-1026.

[12] Bernadó, P; Bertoncini, CW; Griesinger, C; Zweckstetter, M; Blackledge, M. Defining long-range order and local disorder in native alpha-synuclein using residual dipolar couplings. *J. Am. Chem. Soc.* 2005, 17968-17969.

[13] Bertoncini, CW; Jung, Y-S; Fernandez, CO; Griesinger, C; Jovin, TM; Zweckstetter, M. Release of long-range tertiary interactions potentiates aggregation of natively unstructured α-synuclein. *Proc. Natl. Acad. Sci. USA* 2005, 102, 1430-1435.

[14] Svergun, DI; Koch, MHJ. Small-angle scattering studies of biological macromolecules in solution. *Rep. Prog. Phys.* 2003, 66, 1735-1782.

[15] Koch, MHJ; Vachette, P; Svergun, DI. Small-angle scattering: A view on the properties, structures and structural changes of biological macromolecules in solution. *Quart. Rev. Biophys.* 2003, 36, 147-227.

[16] Svergun, DI; Koch, MHJ. Advances in structure analysis using small-angle scattering in solution. *Curr. Opin. Struct. Biol.* 2002, 12, 654-660.

[17] Petoukhov, MV; Svergun, DI. Analysis of X-ray and neutron scattering from biomolecular solutions. *Curr. Opin. Struct. Biol.* 2007, 17, 562-571.

[18] Doniach, S. Changes in biomolecular conformation seen by small angle X-ray scattering. *Chem. Rev.* 2001, 101, 1763-1778.

[19] Guinier, A. La diffraction des rayons X aux tres petits angles: Application a l'étude de phenomenes ultramicroscopiques. *Ann. Phys. (Paris)* 1939, 12, 161-237.

[20] Calmettes, P; Durand, D; Desmadril, M; Receveur, V; Smith, JC. How random is a highly denatured protein?. *Biophys. Chem.* 1994, 53, 105-114.

[21] Feigin, LA, Svergun, DI. Structure analysis by small-angle X-ray and neutron scattering. New York: Plenum Press, 1987, 335 pp.

[22] Flory, PJ. Principles of polymer chemistry. Cornell University Press. Ithaca, NY. 1953.

[23] LeGuillou, JC; Zinn-Justin, J. Critical exponents for the n-vector model in three dimensions from field theory. *Phys. Rev. Lett.* 1977, 39, 95-98.

[24] Kohn, JE; Millet, IS; Jacob, J; Zagrovic, B; Dillon, TM; Cingel, N; Dothager, RS; Seifert, S; Thiyagarajan, P; Sosnick, TR; Hasan, MZ; Pande, VS; Ruczinski, I; Doniach, S; Plaxco, KW. Random coil behaviour and the dimensions of chemically unfolded proteins. *Proc. Natl. Acad. Sci. USA*. 2004, 101, 12491-12496.

[25] Wilkins, DK, Grimshaw, SB; Receveur, V; Dobson, CM; Jones, JA; Smith, LJ. Hydrodynamic radii of native and denatured proteins measured by pulse field gradient NMR techniques. *Biochemistry*, 1999, 38, 16424-16431.

[26] Zhou, HX. Dimensions of denatured protein chains from hydrodynamic data. *J. Phys. Chem. B*, 2002, 106, 5769-5775.

[27] Goldenberg, DP. Computational simulation of the statistical properties of unfolded proteins. *J. Mol. Biol.* 2003, 326, 1615-1633.

[28] Jha, AK; Colubri, A; Freed, KF; Sosnick, TR. Statistical coil model of the unfolded state: Resolving the reconciliation problem. *Proc. Natl. Acad. Sci. USA.* 2005, 102, 37, 13099-13104.

[29] Tran, HT; Wang, X; Pappu, RV. Reconciling observations of sequence-specific conformational propensities with the generic polymeric behaviour of denatured proteins. *Biochemistry* 2005, 44, 11396-11380.

[30] Kazmisrki, SL; Wong, K-B; Freund, SMV; Tan, YJ; Fersht, AR; Daggett, V. Protein folding from a highly disordered denatured state: The folding pathway of chymotrypsin inhibitor 2 at atomic level. *Proc. Natl. Acad. Sci. USA.* 2001, 98, 4349-4354.

[31] Garcia, P; Serrano, L; Durand, D; Rico, M; Bruix, M. NMR and SAXS characterization of the denatured state of the chemotactic protein CheY: Implications for protein folding initiation. *Prot. Sci.* 2001, 10, 1100-1112.

[32] Hodsdon, ME; Frieden, C. Intestinal fatty acid binding protein: The folding mechanism as determined by NMR studies. *Biochemistry* 2001, 40, 732-742.

[33] Shortle, D; Ackerman, MS. Persistente of native-like topology in a denatured protein in 8 M urea. *Science* 2001, 293, 487-489.

[34] Millett, IS; Doniach, S; Plazco, KW. Toward a taxonomy of the denatured state: Small angle scattering studies of unfolded proteins. *Adv. Protein Chem.* 2002, 62, 241-262.

[35] Fitzkee, NC; Rose, GD. Reassessing random-coil statistics in unfolded proteins. *Proc. Natl. Acad. Sci. USA* 2004, 101, 12497-12502.

[36] Meier, S; Grzsiek, S; Blackledge, M. Mapping the conformational landscape of urea-denatured ubiquitin using residual dipolar couplings. *J. Am. Chem. Soc.* 2007, 129, 9799-9807.

[37] Petrescu, AJ; Receveur, V; Calmettes, P; Durand, D; Desmadril, M; Roux, B; Smith, JC. Small-angle neutron scattering by strongly denatured protein: Analysis using random polymer theory. *Biophys. J.* 1997, 72, 335-342.

[38] Zagrovic, B; Lipfert, J; Sorin, EJ; Millett, IS; van Gunsteren, WF; Doniach, S; Pande, VS. Unusual compactness of a polyproline type II structure. *Proc. Natl. Acad. Sci. USA* 2005, 102, 11689-11703.

[39] Bernadó, P; Blanchard, L; Timmins, P; Marion, D; Ruigrok, RWH; Blackledge, M. A structural model for unfolded proteins from residual dipolar couplings and small-angle X-ray scattering. *Proc. Natl. Acad Sci. USA.* 2005, 102, 17002-17007.

[40] Tran, HT; Pappu, RV. Toward an accurate theoretical framework for describing ensembles for proteins under strongly denaturing conditions. *Biophys. J.* 2006, 91, 1868-1886.

[41] Bernadó, P; Mylonas, E; Petoukhov, MV; Blackledge, M; Svergun, DI. Structural characterization of flexible proteins using Small-angle X-ray scattering. *J. Am. Chem. Soc.* 2007, 129, 5656-5664.

[42] Bernadó, P; Pérez, Y; Svergun, DI; Pons, M. Structural characterization of the active and inactive states of Src kinase in solution by Small-angle X-ray scattering. *J. Mol. Biol.* 2008, 376, 492-505.

[43] Wang, S; Overgaard, MT; Hu, YX; McKay, DB. The *Bacillus subtilis* RNA helicase YxiN is distended in solution. *Biophys. J.* 2008, 94, L01-3L.

[44] Kuzmenkina, EV; Heyes, CD; Nienhaus, GU. Single-molecule Föster resonance energy transfer study of protein dynamics under denaturing conditions. *Proc. Natl. Acad. Sci. USA* 2005, 102, 15471-15476.

[45] Hoffman, A; Kane, A; Nettels, D; Hertzog, DE; Baumgärtel, P; Lengefeld, J; Reichardt, G; Horsley, DA; Seckler, R; Bakajin, O; Schuler, B. Mapping protein collapse with single-molecule fluorescence and kinetic synchrotron radiation circular dichroism spectroscopy. *Proc. Natl. Acad. Sci. USA* 2007, 104, 105-110.

[46] Svergun, DI; Barberato, C; Koch, MHJ. CRYSOL – a program to evaluate X-ray solution scattering of biological macromolecules from atomic coordinates. *J. Appl. Cryst.* 1995, 28, 768-773.

[47] Lovell, SC; Davis, IW; Arendall, WB; de Bakker, PIW; Word, JM; Prisant, MG; Richardson, JS; Richardson, DC. Structure validation by Cα geometry: ϕ, ψ and Cβ deviation. *Proteins*, 2003, 50, 437-450.

[48] Dames, SA; Aregger, R; Vajpai, N; Bernadó, P; Blackledge, M; Grzesiek, S. Residual dipolar couplings in short peptides reveal systematic conformational preferences of individual amino acids. *J. Am. Chem. Soc.* 2006, 128, 13508-13514.

[49] Cho, MK; Kim, HY; Bernadó, P; Fernández, CO; Blackledge, M; Zweckstetter, M. Amino acid bulkiness defines the local conformations and dynamics of natively unfolded alpha-synuclein and tau. *J. Am. Chem. Soc.* 2007, 129, 3032-3033.

[50] Mukrash, MD; Markwick, P; Biernat, J; Bergen, M; Bernadó, P; Griesinger, C; Mandelkow, E; Zweckstetter, M; Blackledge, M. Highly populated turn conformations in natively unfolded tau proteins identified from residual dipolar couplings and molecular simulation. *J. Am. Chem. Soc.* 2007, 129, 5235-5243.

[51] Uversky, VN. Natively unfolded proteins: A point where biology waits for physics. *Prot. Sci.* 2002, 11, 739-756.

[52] Segel, DJ; Fink, AL; Hodgson, KO; Doniach, S. Protein denaturation: A small-angle X-ray scattering study of the ensemble of unfolded states of cytochrome c. *Biochemistry* 1998 37, 12443-12451.

[53] Chiti, F; Taddei, N; Bucciantini, M; White, P; Ramponi, G; Dobson, CM. Mutational analysis of the propensity for amyloid formation by a globular protein. *EMBO J.* 2000, 19, 1441-1449.

[54] Bucciantini, M; Giannoni, E; Chiti, F; Baroni, F; Formigli, F; Zurdo, J; Taddei, N; Ramponi, G; Dobson, CM; Stefani, M. Inherent toxicity of aggregates implies a common mechanism for protein misfolding diseases. *Nature*, 2002, 416, 507-511.

[55] Akiyama, S; Takahashi, S; Kimura, T; Ishimori, K; Morishima, I; Nishikawa, Y; Fujisawa, T. Conformational landscape of cytochrome c folding studied by microsecond-resolved small-angle x-ray scattering. *Proc. Natl. Acad. Sci. USA* 2002, 99, 1329-1334.

[56] Uzawa, T; Akiyama, S; Kimura, T; Takahashi, S; Ishimori, K; Morishima, I; Fujisawa, T. Collapse and search dynamics of apomyoglobin folding revealed by submillisecond observations of α-helical content and compactness. *Proc. Natl. Acad. Sci. USA* 2004, 101, 1171-1176.

[57] Kimura, T; Uzawa, T; Ishimori, K; Morishima, I; Takahashi, S; Konno, T; Akiyama, S; Fujisawa, T. Specific collapse followed by slow hydrogen-bond formation of β-sheet in the folding of single-chain monellin. *Proc. Natl. Acad. Sci. USA* 2005, 102, 2748-2753.

[58] Dyson, HJ; Wrigth PE. Unfolded proteins and protein folding studied by NMR. *Chem. Rev.* 2004, 104, 3607-3622.

[59] Redfield, C. Using nuclear magnetic resonante spectroscopy to study molten globule states of proteins. *Methods* 2004, 34, 121-132.

[60] Meier, S; Blackledge, M; Grzesiek, S. Conformational distributions of unfolded polypeptides from novel NMR techniques. *J. Chem. Phys.* 2008, 128, 052204.

[61] Gillespie, JR; Shortle, D. Characterization of long-range structure in the denatured state of staphylococcal nuclease. I. Paramagnetic relaxation enhancement by nitroxide spin labels. *J. Mol. Biol.* 1997, 268, 158-169.

[62] Lindorff-Larsen, K; Kristjansdottir, S; Teilum, K; Fiebre, W; Dobson, CM; Poulsen, FM; Vendruscolo, M. Determination of an ensemble of structures representing the denatured state of the bovine acyl-coenzyme A binding protein. *J. Am. Chem. Soc.* 2004, 126, 3291-3299.

[63] Dedmon, MM; Lindorff-Larsen, K; Christodoulou, J; Vendruscolo, M; Dobson, CM. Mapping long-range interactions in α-synuclein using spin-label NMR and ensemble molecular dynamics simulations. *J. Am. Chem. Soc.* 2005, 127, 476-477.

[64] Marsh, JA; Neale, C; Jack, FE; Choi, WY; Lee, AY; Crowhurst, KA; Forman-Kay, JD. Improved structural characterization of the drkK SH3 domain unfolded state suggest a compact ensemble with native-like and non-native structures. *J Mol. Biol.* 2007, 367, 1494-1510.

In: Protein Misfolding
Editors: C. B. O'Doherty, A.C. Byrne

ISBN: 978-1-60456-881-3

Short Communication B

PREVENTION OF PROTEIN MISFOLDING AND AGGREGATION BY THE MACHINERY OF MOLECULAR CHAPERONES

Carlos H.I. Ramos
Department of Organic Chemistry, Institute of Chemistry
University of Campinas-UNICAMP
P.O. Box 6154, 13083-970, Campinas, SP, Brazil

ABSTRACT

Amyloid fibrils and plaques that form deposits are characteristic of several degenerative conditions affecting a large variety of tissues. Almost constantly new evidences are linking protein misfolding and aggregation to one or more type of diseases, leading to the hypothesis that a considerable part of the known diseases are caused by the increase of these non-functional molecules. Molecular chaperones are a family of proteins initially identified as heat shock proteins that are involved with assistance of the correct folding and prevention of aggregation of unfolded polypeptides. The constitutive induction and the housekeeping characteristics of these proteins were recognized later. In addition to that, recent works have shown that these proteins are also capable of disaggregation and refolding of stable protein aggregates, a novel function so spectacular that raised the hypothesis that molecular chaperones may be as important in regulating the malefic effects of misfolded or aggregated proteins as in raising the yield of functional molecules. These new findings have also led to the hypothesis that the functional failure of the molecular chaperone machinery is a potential trigger of degenerative conditions. A functional failure that would be more common as the organism ages favoring degenerative diseases in the elderly. In conclusion, new evidences on the function of molecular chaperones have enhanced the potential of these proteins as targets for therapeutic approaches aimed in decreasing the toxicity of aggregates.

1. Introduction

The subject of molecular chaperones is vast and would need a series of books to be covered. However, the number and quality of research that is currently being done with these proteins is so intense that a considerable part of such imaginary series will be obsolete soon. Therefore, the goal of this chapter is to present general information about molecular chaperones and in particular on its fantastic action of preventing protein misfolding and aggregation. For that the chapter has the following structure: 1) A condensed introduction on protein folding and misfolding covering few aspects necessary to understand the action of chaperones. Detailed aspects of misfolded proteins are subjects of other chapters of this book. 2) A general presentation of the main characteristics of molecular chaperones with special attention to the major families and their presence in the cell. and 3) Specific relevant information on the structure and function of the main molecular chaperone families followed by the description of a third aspect, named here Therapy, that will deal with the presentation of results that illustrate the involvement of chaperones with cellular malfunction and attempts to cure related diseases.

The goal of this chapter is to overview the action of chaperones on the prevention of aggregation and disaggregation of misfolded proteins. I tried to cite important works and in some cases a review that cite important works and apologize in advance for eventual absences due to space constrain. Some works are described in more detail aiming to provide the reader with important historical perspective and/or technical angle.

2. Protein Folding, Misfolding and Aggregation

It is generally accepted that all the information necessary for a protein to fold to its native conformation is in its amino-acid sequence (for a review see Ramos & Ferreira, 2005). However, efficient reversible folding and unfolding has been usually observed only for small proteins. For many proteins, refolding experiments led to partially unfolded, misfolded molecules or aggregation, which is unproductive because the function of most proteins is ordinarily related to its native conformation[1]. Seminal works from Dobson´s group (Fandrich et al., 2001; Fandrich & Dobson, 2002) showed that any protein sequence have the potential to form amyloid fibrils given the appropriate conditions. Such findings suggested that not only the native structure and well-folded intermediates are free-energy minima in the landscape of conformations but the amyloid fold could also be another free-energy minimum of all polypeptide chains. The major pathological effect of the inappropriately folded proteins is the deleterious effect of aggregates and the tissue deposition of protein aggregates *in vivo* causes degenerative diseases and organ damage (Dobson, 2004): Examples of conformational diseases are cystic fibrosis, phenylketonuria, short-chain acyl-CoA dehydrogenase deficiency, Alzheimer, Parkinson, prion and some cancers.

---

[1] That is not always the case. The existence of Intrinsically Unfolded Proteins, or IUP, is now well accepted. For the reader eager for knowing more I suggest the following review and references therein: Dyson and Wright, 2006.

Protein aggregation is primarily driven by unspecific hydrophobic interactions, a condition favored under physiological conditions where the very high concentration of macromolecules decreases the excluded volume. Macromolecules excluding volume from each other leads to restriction of configurational entropy and results in the stabilization of oligomers in crowded solutions (for more information see Minton, 2000). Therefore, the tendency that some partially folded chains, such those that cause human amyloid diseases, have to aggregate will increase by crowding inside the cell. This condition may have terrible consequences because aggregates have the property of becoming infections by promoting aggregation of native conformers as the case with prions (Prusiner, 2001).

Some researchers have recently pointed out that misfolded proteins would be in the origin of as much as half of the human diseases (see Chaudhuri & Paul, 2005 and references therein). In this view, a complex formed by molecular chaperones and the proteassome (responsible for protein degradation) would constitute the quality control system that deals with proper protein folding and degradation. Moreover, an accumulation of misfolded proteins with time would overload the chaperone system resulting in the increase in the number of diseases verified in the elderly (Csermely, 2001). Thus cellular components such as chaperones and proteases, as well as environmental factors, may combine to form a complex system that modulates the phenotype of disorders caused by protein misfolding. In conclusion, the comprehension of this system may advance the finding of new therapeutic approaches.

The formation of aggregation is worsened in stress conditions, such as an increase in temperature. An anti-stress cellular response is the expression of factors that protect proteins from unfolding and aggregation. Molecular chaperones are the major factors that had enhanced expression during stress although many of them are also expressed constitutively and have important housekeeping functions. Proteins that escape the initial action of chaperones and precipitate can be ressolubilized by other chaperones. This is important because in a major event in which a stress condition is present the limited chaperone conditions may not be sufficient to deal with all proteins. Therefore, molecular chaperones constitute the central cellular defense against protein misfolding and aggregation that have major pathological consequences (Dobson, 2003). Additional information, which has a substantial meaning, is that chaperones do not influence the final tertiary structure of a protein as it is a characteristic inherent of the amino acid sequence.

3. Molecular Chaperones

3.1. General Characteristics

The folding of many proteins in the cell is assisted by molecular chaperones that are also referred to as heat-shock proteins (Hsp) for being first described as proteins induced during thermal stress (There are many reviews on this subject, to name a few Beissinger and Buchner, 1998; Hartl and Hayer-Hartl, 2002; Borges and Ramos, 2005). The concept of molecular chaperones is better described by Hendrick and Hartl (1993) statement: "*we define a molecular chaperone as a protein that binds to and stabilizes an otherwise unstable conformer of another protein, facilitates its correct fate in vivo: be it in folding, oligomeric assembly, transport to a particular subcellular compartment, or controlled switching between active/inactive conformations*". However, the term molecular chaperone was used first by Laskey and co-workers (Laskey et al, 1978) to describe the function of nucleoplasmin, a protein that promotes the correct oligomerization of histone by preventing its aggregation during assembly in chromatin.

In a general way, molecular chaperones help other proteins to fold which give them a housekeeping function and an administrative property as a result of their capability of activating cellular processes by regulating the folding/function of key proteins. For instance, when cells are submitted to an increase in temperature, mRNAs related to Hsps are shortly detected after heat shock. Molecular chaperones have also a prophylactic action because they prevent proteins with aggregative potential to form amyloid fibrils. In addition to that, molecular chaperones have the ability to dissolubilize aggregates, a powerful characteristic that open the possibility to use them as therapy against diseases caused by misfolded proteins involved in diseases. As a consequence of these properties, molecular chaperones integrate a major central hub involved with the cellular quality control process. A medical application for molecular chaperones is easy to visualize but an agricultural relevance has also been recently raised because significant global warming would affect plants badly since they are sessile and thus more exposed to changes in the environment.

The deleterious action of protein misfolding causes both loss of function, as in the case of some cancer diseases, and gain of damaging function, as in the case of Alzheimer and Parkinson diseases. The ultimate goal will be the developing of small molecules that manipulates chaperones and protein homeostasis. As an example, compounds that could induce a high expression of molecular chaperones would be important as therapeutic agents to increase the yield of correctly folded proteins and to decrease the number of aggregates in pathological conditions. The heat shock response (HSR) is transcriptionally regulated by a heat shock factor (HSF) that triggers the expression of molecular chaperones. Some compounds that increase the action of HSF are in clinical trials and had positive impact on oxidative stress and on some cancers (see Macario and de Macario, 2007 and references therein).

3.2. Categorization

Members of molecular chaperones are categorized in five major families which are classified by their molecular mass (Figure 1 and Table 1). Each family usually has more than one sub-family that is usually classified in accord to cellular localization, function and/or expression pattern. These chaperones are assisted by other molecular chaperones known as co-chaperones which exist in higher number than chaperones. As a matter of fact, the number of chaperones does not have significant change when single organisms are compared with complex ones, while the number and variety of co-chaperones change significantly with organism complexity. This important characteristic may be necessary for the quality control system to cope with the complex intracellular organization of superior organisms. Due to the large number and variety, just a few co-chaperones will be briefly described in this chapter.

The exception will be the co-chaperones from the Hsp40 family because they are highly important for many cellular activities and are involved in helping several chaperones with crucial importance for cellular quality control. In fact, many researches consider that co-chaperones are sorted in two classes based on the structure of their domains: the 'J domain' and the 'TRP domain' classes. As the names reveals the first is characterized by the present of a J domain (mostly Hsp40s) and the later by the presence of a tetratricopeptyde (TRP) domain that is usually present in co-chaperones that interacts with Hsp70 and Hsp90.

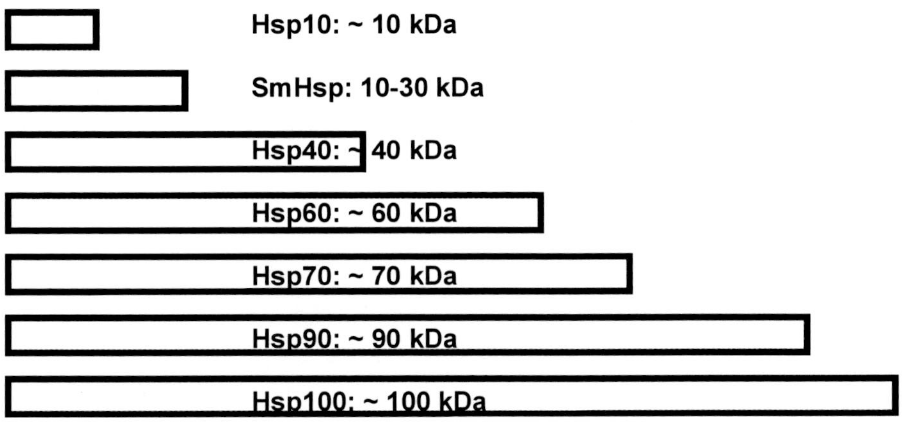

Figure 1. Molecular mass and relative sizes for molecular chaperones. The values are for monomers: Hsp 10 is a heptamer, SmHsp is 8-24mer, Hsp40 is a dimer, Hsp60 is a double heptamer, Hsp70 is a monomer, Hsp90 is a dimer and Hsp100 is a hexamer. Hsp40 is a co-chaperone.

Table 1. A summary of general characteristics of molecular chaperones

GENERAL NAME	MOLECULAR MASS[a]	FUNCTIONAL OLIGOMER	*E. coli* GENE	YEAST GENE
Hsp100	~100	Hexamer	*clpB*	*hsp104*
Hsp90	~90	Dimer	*htpG*	*hsp82*
Hsp70	~70	Monomer	*dnaK*	*ssa, ssb*
Hsp60/Hsp10	~60/~10	Double heptamer/ heptamer	*groEL/groES*	*hsp60/hsp10*
Small Hsp	10-30	8-24	*ibpA,ibpB*	*smHsp*
Hsp40[b]	~40	Dimer	*dnaJ*	*ydj1, sis1*

[a], for the monomer, kDa; [b], Hsp40 is a co-chaperone.

The five main families of molecular chaperones are Hsp70, Hsp60/Hsp10, Hsp90, Hsp100 and smHsp (Figure 1 and Table 1). Table 1 lists their molecular mass and oligomeric state. Table 1 also lists the *E. coli* and yeast name of these chaperones because of the much work that has been done with them. A brief and general description of each family follows. Hsp70 has a central role in the molecular chaperone system because it is involved with the folding of nascent proteins and in the intercommunication between substrate and chaperones in the cell. Its importance can be recognized by the fact that each organelle has its own Hsp70 protein. Hsp70 has ATPase activity and has two domains, one that binds substrate and other that binds nucleotide. Hsp60/Hsp10 form a barrel-like structure within partially unfolded proteins can find the proper environment to fold avoiding aggregation. The Hsp60/Hsp10 complex also depends on the hydrolysis of ATP to work. Hsp90 are essential chaperones to the cell and are mainly involved with regulation by helping the folding of many factors involved in signal transduction. Hsp90 works as a dimer and also binds ATP. Hsp100, another chaperone with ATPase activity, form ring-like high oligomers and has the important function of disaggregating proteins. One of the hypotheses for its function is that it works as a crowbar pulling polypeptides from the aggregates and delivering one by one to Hsp70 for proper folding. Small Hsps or SmHsps had this name because the monomers have low molecular masses when compared with other chaperones. However they formed oligomers with several subunits, usually 8 to 30. They do not have ATPase activity and work by binding unfolding proteins during heat shock protecting them from aggregation. They are also involved in delivering the substrates to Hsp70 for proper folding. It is possible to separate molecular chaperones in 'foldases', which are mainly involved in folding proteins as is the case of Hsp70 and Hsp60/Hsp10, and 'holdases', which are mainly involved with protecting partially folded polypeptides from aggregation and delivering, as is the case of smHsps and the co-chaperone Hsp40. In some conditions chaperones can have both holdase and foldase functions.

3.3. Genes and Expression

The importance of molecular chaperones can be confirmed by the number of genes belonging to them present in the genome and by the high expression of these genes. Plants are sessile and therefore exposed to all kind of environmental conditions. Extreme conditions may cause stresses situations where the expression of molecular chaperones is of great significance for survival. The genome of the model plant *Arabdopsis thaliana* has been sequenced and the gene expression profile, measured by RNA expression, of many plants, among them sugarcane and eucalyptus, have been investigated. The data provided by these works can be mined specifically for molecular chaperones and used to generate new information about them (The TIGR *Arabidopsis thaliana* Database: http://www.tigr.org/tdb/e2k1/ath1/; Borges et al., 2001; Cagliari et al., 2005). About 2% of all expressed genes in sugarcane are molecular chaperones which gives an appropriate dimension of the importance of these proteins for the cell. From these, about 20% belong to the Hsp70 family, about 20% belong to Hsp70 co-chaperones and about 10% belong to the Hsp90 family (Table 2). This result emphasizes the importance of Hsp70 and Hsp90 in regulating many cellular processes. Cytosolic Hsp70, Hsc70, form with Hsp90 and co-chaperones a multifunctional chaperone machinery that protects against genetic and environmental stress. Such machinery has no correspondent in prokaryotes.

Table 2 describes the percentage of mRNA expression for the main molecular chaperones described here. The values annotated for eucalyptus correlate very well with those annotated for sugarcane indicating that they may have a parallel expression level and type of response to stress. In order to expand the results for a global behavior in plants would request more experimental studies. However, it is interesting to note that the number of genes belonging to molecular chaperones in sugarcane correlates well with those in *Arabdopsis thaliana* (Table 2).

Table 2. A simple genomic analysis of molecular chaperones using three plant representatives. *Arabidopsis thaliana* genes where mined using the word *hsp* as search query in: http://www.tigr.org/tdb/e2k1/ath1/, results from sugarcane and eucalyptus are published elsewhere (Borges et al, 2001-; Cagliari et al., 2005-).

Molecular Chaperone	*Arabidopsis thaliana* (number of genes)	Sugarcane (predicted number of genes)	Sugarcane (expressed tags) [a]	Eucalyptus (expressed tags) [a]
Hsp100	5	9	3%	7%
Hsp90	3	6	10%	18%
Hsp70	14	8	20%	25%
Hsp60	2	10	8%	12%
Small Hsp	19	11	5%	4%
Hsp40[c]	-	39	22%	28%

[a], percentage of expressed molecular chaperones (which are about 2% of all expressed cDNAs); [b], Hsp40 is a co-chaperone.

4. Structure, Function and Therapy

4.1. HSP70

4.1.1. Struture and Function

Hsp70 is the most highly conserved protein in evolution and eukaryotic family members are localized in the cytosol, endoplasmic reticulum and mitochondria (Craig, 1989) and constitutes the central part of a ubiquitous chaperone system that is the major machinery utilized to protect cells from stress and facilitate cellular protein folding (Mayer *et al.*, 2001; Hartl and Hayer-Hartl, 2002; Wegele *et al.*, 2004). Eukaryotes have more than one gene encoding Hsp70 proteins and the human Hsp70 family consists of at least eight gene products with individual cellular locations, with specific function and with either stress- or constitutive-induction (Daugaard et al, 2007). It is worth mentioning the Hsp70-1a chaperone which is located in the cytosol, nucleus and lisossome and is heat-induced and Hsc70 which is located in the nucleus and citossol and is constitutively expressed. There is also a mitochondrial Hsp70, Hsp70-mt, which is highly similar to *E. coli* DnaK, and BiP (for Binding Protein) which is the specialized Hsp70 found in the endoplasmic reticulum and has an important role in the unfolded response. Hsp70 has ATPase activity which is enhanced by Hsp40 proteins but the precise affinity of Hsp70 for either ATP or ADP is still on debate. Reported K_D values for DnaK is in the order of nM for both ATP and ADP and reported K_D

values for mammal Hsc70s range from nM to μM for both ATP and ADP (Theyssen et al., 1996; Borges and Ramos, 2006).

All types of Hsp70 have a two-domain structure, an N-terminal ATPase domain of about 45 kDa (NBD for Nucleotide Binding Domain) and a C-terminal polypeptide-binding domain of about 25 kDa (SBD for Substrate Binding Domain), which is critical for the regulated interactions of this chaperone (Figure 2). There is also an EEVD-motif at the C-terminus that is responsible to interact with other chaperones (Figure 2A). Hsp70 utilizes its polypeptide-binding domain to bind non-native polypeptides preferentially to motifs composed of four-five hydrophobic residues flanked by basic residues. The ATPase domain regulates the affinity for substrates because the hydrolysis of ATP causes a conformational change in the ATPase domain that is transmitted to the polypeptide binding site which modulates the affinity for unfolded substrates. Several experimental evidences suggest that the domains are moving in solution as a single unit and not as two individual structures connected by a flexible linker (Figure 2B), i.e. there is allostery caused both by nucleotide and peptide binding. In this mechanism the conformational change caused by ATP binding would propagate throughout a flexible linker that is 10 amino acids longer and connects the NBD to the SBD. When ATP is bound the linker becomes more sensitive to proteases (Jiang et al, 2005), i.e. it is moving from a less exposed to a more exposed position.

ATP binding is important because the affinity for client proteins is increased by several orders of magnitude upon hydrolysis of ATP to ADP and inorganic phosphate. On the other hand, substrate binding to the SBD stimulates ATP hydrolysis. However, Hsp70s in general have only low ATPase activity and the action of Hsp40 or Hsp40-like co-chaperones, which stimulate ATP hydrolysis, is required. We believe that Hsp40s proteins form complexes with unfolded proteins to prevent their aggregation and deliver them to Hsp70. The stable Hsp70-substrate complex is formed by a mechanism that involves the binding of the Hsp40 J-domain, which will stimulate the hydrolysis of ATP.

Evidences for the role that Hsp70 plays during the initial phases of protein translation came from two main observations: 1) Hsp70 is associated with ribosome-bound nascent chains and 2) mitochondrial and endoplasmic Hsp70s are associated with translocating polypeptides. Hsp70s bind short linear peptide segments with a net hydrophobic character, while it is enough to promote the folding of many proteins *in vitro*, in many other cases the Hsp70-bound substrate has to be transferred to a chaperonin complex for productive folding. In a clever series of experiments Hartl´s group (Teter et al., 1999) use the antibiotic puromicyn to release *in vivo* nascent polypeptides from the ribosome. These polypeptides had puromicyn bound to its C-terminus and could be probed by anti-puromicyn antibodies. The complexes were separated by centrifugation and tested for the presence of DnaK. They found that ~20% of the proteins found in these experiments were bound to DnaK and that the client proteins ranged from 16-80 kDa but with notable preference from chains of 4-65 kDa. When a Δtig (for Trigger factor, TIG) vector was tested using the same experimental approach a two-fold increase in nascent polypeptides associated with DnaK was detected and many of them were short chains. The conclusion is that TIG binds most of the short nascent polypeptides *in vivo* that usually will not bind to DnaK. The cooperation between TIG and DnaK seem to be important since Δtig and ΔdnaK but not the double Δtig-ΔdnaK deleted mutant is viable (see below).

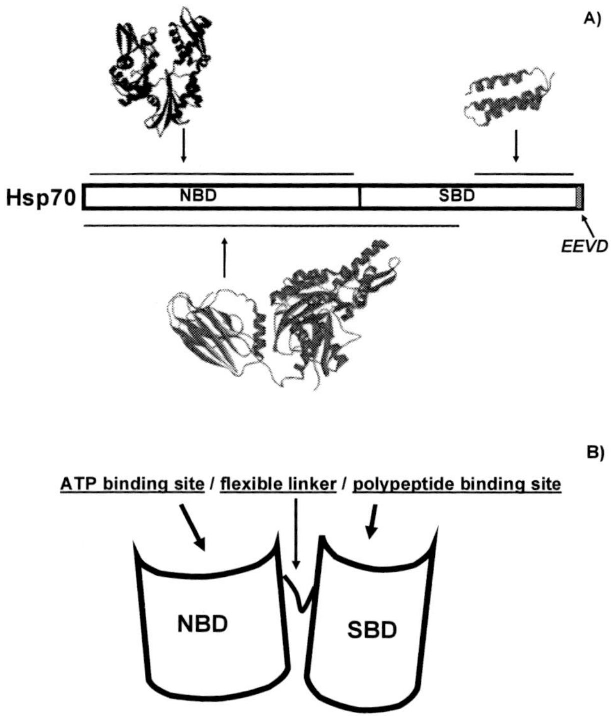

Figure 2. Schematic representation of Hsp70. A) Top black: human Hsp70.1 residues 2-382, PDB accession number 1S3X. Top gray: *C. elegans* residues 533-614, PDB accession number 2P32. Bottom gray: bovine (*Bos taurus*) Hsc70 residues 1-554, PDB accession number 1YUW. The EEDV sequence is indicated in gray. Figure composed using Weblab ViewerLite 3.7 from Molecular Simulations Inc.B) The Nucleotide Binding Domain (NBD), the Substrate Binding Domain (SBD) and the ~10 kDa linker are shown.

As described above Hsp70 were about 20% of all molecular chaperones expressed in eucalyptus. Hsp70 homologues from all cellular compartments, but the cytoplasmic Hsp70 (Hsc70 subfamily), represented more than half of all reads related to Hsp70 family, which is about 10% of all reads related to chaperones. Hsc70 represented also more than half of Hsp70 proteins in sugar cane cells and plays an important role in the net response to low temperature in sugar cane (Nogueira *et al.*, 2003). These findings point to a universal pattern of expression for Hsp70 in plants. A rough estimation suggests that Hsp70 genes alone are responsible for almost 1% of the entire gene expression in plant cells.

4.1.2. Importance for Thermostability and Client Proteins

An important experiment showing that Hsp70 is the single most effective chaperone in preventing protein aggregation, as well as identifying his thermolabile client proteins, have been done using *E. coli* proteins (Mogk et al., 1999). Soluble *E. coli* extracts were pre-incubated at 30°C for 5min in the absence and in the presence of chaperones and ATP was added 2 min prior heat-shock. Then samples were shifted to 45°C for 15min and aggregated proteins were pelleted by centrifugation and analyzed by one dimension- and two dimension-gel electrophoresis. The results showed that 10-15% of the proteins precipitated in these conditions regarding the presence of endogenous chaperones. At heat-shock conditions the most expressed chaperones found were DnaK and GroEL and the chaperone that better suppressed aggregation was DnaK which reached a 90% efficiency when added at physiological concentrations. The proteins protected by DnaK were identified by mass spectroscopy methods and a list is presented in the paper. The proteins corresponded to ~30% of the proteins detected by the experiment and those that are clients of DnaK range in size from 16 to 151 kDa, have a large variety of cellular function, and are essential for cell viability.

The authors have also used knocked-out mutants of *E. coli* in which molecular chaperones were deleted (ΔdnaK, ΔclpA, ΔclpB, Δhtg and ΔibpAB) and analyzed cell standardly and heated-shock treated cells by two dimension-gel electrophoresis. They were able to show an increasing in protein aggregation on heat–stress and were able to identify protein-clients by comparing the profile from wild-type and knocked-out cells. Significant increase in protein aggregation was identified only in the ΔdnaK compared to the other strains. The controlled overexpression of other chaperones, as GroEL, HtpG or ClpB, was not able to revert the phenotype of the knocked-out cells. Only a massive overproduction of GroEL/GroES (~30% of total protein) was able to only partially suppress protein aggregation. These results lead to the conclusion that Dnak is the most effective chaperone system to prevent protein aggregation in the cell.

4.1.3. Therapy

Connections between the inhibition of Hsp70 and the rising of several diseases and also between a stimulation of Hsp70 in cells over stress have been widely observed and studied. 1) Neuronal cells from rats show an increase in Hsp70 when exposed to hiperthermic conditions. 2) The aggregation of alpha-sinuclein, a protein involved with Parkinson disease, is inhibited by Hsp70. 3) The concentration of Hsp70 in cells with mutant huntingtin, involved with Huntington disease, is higher than in cells with the wild-type protein. 4) A combination of Hsp70, Hsp90 and Hsp40 prevents the aggregation of the β-amiloyd peptide involved with Alzheimer.

Much of what is known about neurodegenerative diseases came from the studies with *Drosophila* which has become an excellent model for this kind of study. In this model genes associated with Huntington, caused by polyglutamine repeat proteins, and Parkinson, caused by α-synuclein, diseases can be expressed generating phenotypes similar to that of the human diseases. Using this system Bonini (2002) found that Hsp70, the main stress-induced molecular chaperone in flies, coimmunoprecipitated with aggregated polyglutamine repeat proteins and enhanced expression of Hsp70 suppressing the deleterious action of these aggregates. She also showed that an Hsp40 protein enhanced the effect of Hsp70 alone and that an Hsp70 mutant with low ATPase activity was unable to suppress degeneration. Similar results were found when flies for Parkinson disease model were studied.

Tumoral cells have high levels of Hsp70 mRNAs and Hsp70 is involved in the regulation of apoptosis. Hsp70 binds in a stable manner to antigenic peptides in the cytosol of tumor cell and this complex is released when the cell dyes and are recognized by receptors of immune cells. Therefore Hsp70 is the basis for immunotherapy protocols because it helps antigen presentation and breakes self-tolerance. To sum up, compounds that are able to activate the function of Hsp70 are pharmaceutical targets and raising either the ATPase activity or the interaction between Hsp70 and Hsp40 is an important aspect that can be stimulated by compounds aimed to improve Hsp70 activity.

4.1.4. Nascent Polypeptides and Trigger Factor

The fate of nascent proteins in eubacteria does not depend only on Hsp70 but also on trigger factor (TIG ou TF, ~48kDa). Apparently TIG is only present in eubacteria since high identity between TIG and ribosome binding proteins in other organisms have not been identified yet. TIG is found in excess when compared to the number of ribosomes present in the cell. Trigger factor bound to ribosome in a 1:1 stoichiometry but when free in the citossol is present as dimers. A hypothesis to the mechanistic action of this protein suggests that it undergoes conformational changes when binding to the ribosome resulting in exposing a surface to hydrophobic patches in the polypeptide (Figure 3). In this mechanism, TIG (one or more) remains bound to the polypeptide if necessary until DnaK bound to it and may be considered a "holdase' since it does not bind nucleotide.

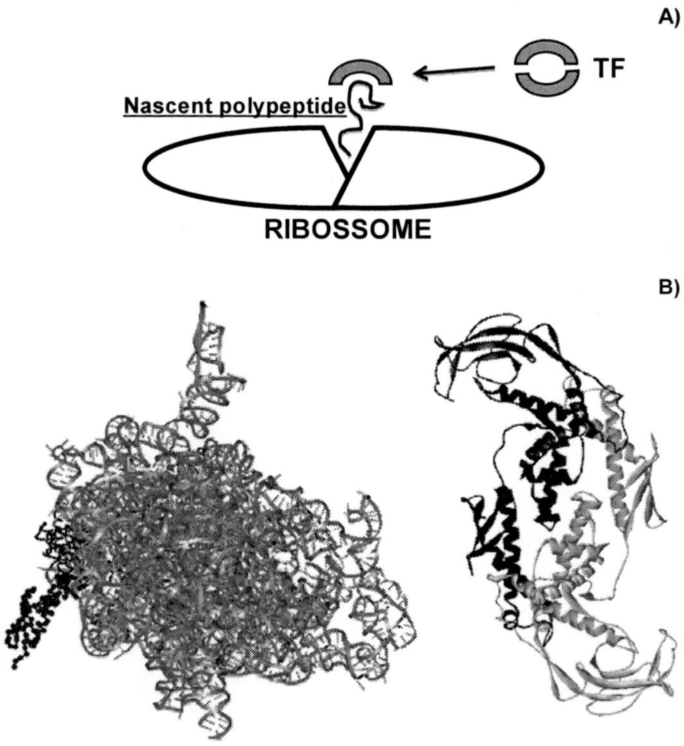

Figure 3. A) Schematic representation of trigger factor (TF) bound to a polypeptide nascent from a ribossome. B) Right: crystal structure of TF associated with ribossome, PDB accession number 2AAR; ray: 23S ribosomal RNA; black tubes: 50S ribosomal proteins L23 and L29; black tube and dots: trigger factor. Left: dimeric structure of TF, PDB accession number 1T11. Figure composed using Weblab ViewerLite 3.7 from Molecular Simulations Inc.

Much of our knowledge about trigger factor is recent and has come from research that has been done in the last decade. One important research was that done with the use of fluorescent probes located in strategically residues (Kaiser et al., 2004). The emitted fluorescence was measured *in vitro* both in the presence and in the absence of ribosome. More than one residue was probed in such a way that the Föster resonant energy transfer (FRET), which can be related to distance, between them could be measured. The results showed that the distance changed in the presence of ribosome, i.e. Trigger factor undergoes conformational changes in the presence of ribosome.

In vivo-like experiments using cell-free translation systems were also performed and they are confirmatory of conformational changes caused by ribosome association. In addition to that, experiments combined with translating radiolabeled Methionine proteins showed that TIG was recruited to translating ribossomes and its conformation shifted from a close to an open state. Competition experiments using unlabeled Trigger factor showed that compaction of its structure occurred preferentially in concomitance with its disengagement from the ribosome. However, the experimental analyses of a range of polypeptides showed that TIG remains bound to polypeptides with high hydrophobicity for longer periods of time and the affinity of the polypeptide for Trigger factor decreases resulting in dissociation once the hydrophobic patches are buried by folding.

Additional experiments showed that Trigger factor remains bound to the nascent polypeptide even after it had completely emerged from the ribosome suggesting that TIG delays folding. It is interesting to note that *E. coli* has a lower rate of multidomain protein folding per ribosome when compared to Eukaryote cells. There is cooperation between Trigger factor and DnaK in *E. coli* favoring post-translational but not co-translational folding. The authors suggest that these characteristics may be responsible for the low solubility yields of recombinant Eukariotic proteins produced in *E. coli* (Kaiser et al., 2004). More experimental study is necessary to determine whether co-expression of other chaperone or co-chaperone, perhaps a 'holdase', could help to improve the folding of recombinant proteins. Anyway, additional information about this system can help the understanding of nascent folding in eukaryotes, either by finding a similar system in the eukaryotic cell or by mimetic solutions to prevent misfolding.

4.2. HSP40

4.2.1. Struture and Function

Proteins belonging to the 40 kDa heat shock (Hsp40) family are co-chaperones that bind specific clients and deliver them to Hsp70s (Cyr et al, 1994; Cheetham and Caplan, 1998), i.e. they act as 'holdases'. In addition to that, Hsp40 proteins regulate ATP-dependent polypeptide binding promoting the formation of a stable complex between Hsp70 and the polypeptide (Liberek et al., 1991). As a result, different Hsp70-Hsp40 pairs binding unique clients can be formed because Hsp40 members have specialized functions. All members of the large Hsp40 protein family contain an N-terminal J domain that is responsible for stimulating Hsp70 ATPase activity. The J domain was first identified in the *Escherichia coli* DnaJ which is comprised of four domains: the J domain, a Gly/Phe-rich (G/F) region, a Cys-rich domain, and a C-terminal region (Figure 4). There are two main classes of Hsp40s that have been classified depending on the presence or absence of the G/F region and the Cysteine repeats. Type I Hsp40-like proteins (e.g. *Escherichia coli* DnaJ, yeast Ydj1, human DJA1) contain the J domain, the G/F region, and the Cysteine repeats, while type II Hsp40s (e.g. yeast Sis1, human DJB4) have the J domain and a G-rich region with one segment rich in Phenilalanine residues (G/F) and a second rich in Methionine residues (G/M). Most of the studies on Hsp40 proteins have been carried out with yeast type I Ydj1 and type II Sis1. Both proteins can stimulate the ATPase activity of Hsp70 Ssa proteins facilitating the refolding of unfolded proteins (Lu and Cyr 1998). However they have distinct chaperone functions: Ydj1 can function independent of Hsp70 to directly bind nonnative proteins and suppress protein aggregation while Sis1 must function with Hsp70 to suppress aggregation. Ydj1 is involved with the translocation of proteins from the cytosol into organelles while Sis1 is found in association with translating ribosomes and is linked to the assembly of translation initiation complexes. Moreover, *sis1* is an essential gene and the lethal phenotype of a sis1 strain is not complemented by Ydj1, however Sis1 can suppress the slow-growth phenotype of an ydj1 strain.

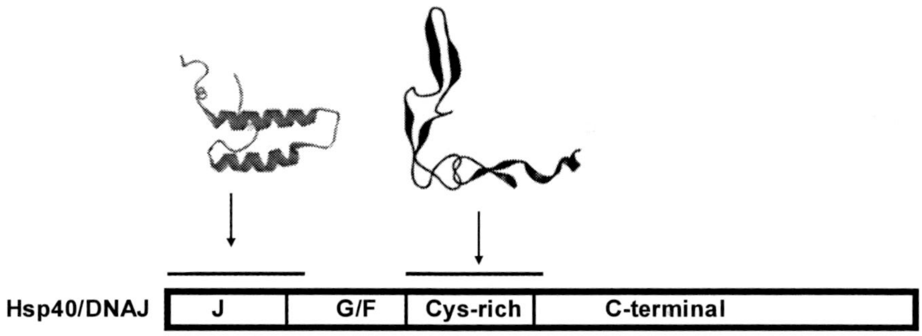

Figure 4. Schematic representation of Hsp40 (DnaJ from *E. coli*). Top gray: J domain residues 2-76, PDB accession number 1XBL. Top black: Cys-rich domain residues 131-209, PDB accession number 1EXK. Figure composed using Weblab ViewerLite 3.7 from Molecular Simulations Inc.

Examination of the domain structures of Ydj1 and Sis1 shows that they have a modular configuration with low similarity at the central domains: Sis1 contain insertions in the G/F region (Lopez et al 2003) and the Cys-rich and G/M regions are not equivalent. To test the hypothesis that these regions are involved in the *in vivo* function, Fan et al (2004) engineered chimeric forms of Ydj1and Sis1 in a way that the chaperone modules were swapped to form YSY and SYS. This study suggests that the chaperone modules of Ydj1 and Sis1 are exchangeable and help to specify Hsp70s cellular functions demonstrating that centrally located protein modules control the protein folding activity of type I and type II Hsp40s. From these studies arose the hypothesis that the difference in function is linked with the structural disposition of the central modules of Hsp40. In accord with this hypothesis studies comparing the low-resolution structures of human type I and type II Hsp40s showed that these proteins have very different quaternary structure: type I Hsp40 form a compact structure in which two highly asymmetrical monomers bound in a bullet shape, while type II Hsp40 had an N-termini with a completely different arrangement in which they pointed away from the compact C termini core responsible for dimerization (Borges *et al.*, 2005). Altogether these results are indicative that type I and type II Hsp40s have different quaternary structure that is regulated by the central modules (Fan et al, 2004; Borges *et al.*, 2005).

4.2.2. Therapy

Several authors have pointed that the Hsp70/Hsp40 system may have therapeutic potential against misfolding diseases. Of particular interest is the study with [URE3], an yeast epigenetic factor that represents the prion form of the autocatalytic conversion of Ure2p into infectious amyloid fibrils. The [URE3] prion phenotype can be cured by the overexpression of Ydj1 (Moriyama et al., 2000) that, different from Sis1, binds directly to Ure2 inhibiting the formation of fibrils (Lian et al., 2007), and overexpression of the Hsp70 Ssa1 protein also cures [URE3] (Schwimmer, 2002). It is clear that Sis1 and Ydj1 have unparalleled function and mode of interaction with Hsp70 Ssa proteins, and investigation of the structural differences may help to understand their dissimilarities. Additional evidence for the combined action of Hsp70 and Hsp40 in preventing aggregation was presented with studies using *E. coli* DnaK and DnaJ (Acébron et al, 2008) showing that a DnaK mutant with low affinity for DnaJ failed to complement ΔdnaK at heat shock temperatures. However, the addition of high concentrations of DnaJ stimulates the mutant to reach wild-type ATP hydrolysis levels *in vitro* and helped the ΔdnaK cell complementation at heat shock temperatures. In support to these findings, studies with *Drosophila* models for Huntginton diseases, expression of polypeptide repeats, showed that the overexpression of Hsp40 proteins seems to suppress the toxicity effect of aggregates. Altogether, studies investigating the action of Hsp40 in preventing protein aggregation point out that this co-chaperone is an important target for approaches aiming to prevent diseases caused by protein misfolding.

4.3. HSP60/HSP10

4.3.1. Struture and Function

Highly studied and equally important is the chaperone system formed by Hsp60 and Hsp10, also known as chaperonin and sometimes referred as 'the Anfisen's cage'. Charles Anfinsen´s work (1961) lead to the hypothesis that all the information for the generation of the native conformation of a protein is contained in its amino acid sequence. It was assumed at that time that *in vivo* folding would proceed in similar manner. However, many evidences, as for instance that of an *E. coli* deficient in an operon named *groE* which were unable to support the growth of wild type phage λ (Georgopoulos et al, 1972), suggested the existence of specialized proteins that participate in the folding of the active structure of others. The identification of the heat inducible Hsp60 as a substrate binder and folding mediator was made in late 1980s by the groups of Hartl, Horwich and Lorimer (see Lorimer, 2001 for a historical overview). GroEL and GroES are indispensable proteins for bacterial growth at temperatures between 20-40°C. *In vivo* observation of polypeptide flux through this chaperonin system showed that they are predominantly between 10-55 kDa and accounted for 10%-15% of all cytoplasmic protein under normal growth conditions, and for 30% or more upon exposure to heat stress (Ewalt et al., 1997). The chaperonin complex is active as a double ring oligomer structure with a cylindrical shape and a mass of almost 1 million Daltons. The chamber is composed of two heptamers of GroEL arranged as two stacked rings with a GroES heptamer at the top acting as a lid (Figure 5). Overproduction of this chaperone system suppresses heat-induced aggregation and is widely used to help the production of heterologous proteins *in vivo*.

Initial understanding on the chaperonin system came from the study of the enzyme RUBISCO, the major synthesized protein in chloroplasts (chaperonins were initially referred to as RUBISCO binding proteins). By combining the use of protein electrophoresis and autoradiography it was possible to isolate RUBISCO. The experiments also showed that during its synthesis RUBISCO was associated with a large complex made of ~60 kDa subunits (see Ellis and Hartl, 1996, for a review). One important experiment that grounded the basis of structure and function was made by Lorimer´s group using a chemically unfolded protein RUBISCO. The unfolded enzyme aggregates if diluted in refolding buffer but not in the presence of purified GroEL, in which they bound with a stoichiometry of 1 enzyme per GroEL double ring. The addition of ATP and GroES to Rubisco-GroEL resulted in the recovering of functional RUBISCO (Goloubinoff et al., 1989). They also showed that GroEL/GroES did not interact with folded RUBISCO and since they used *E. coli* GroEL/GroES in their experiments they concluded that this system should be of general use. They also predicted that this system was able to recognize elements in the unfolded chain binding to them.

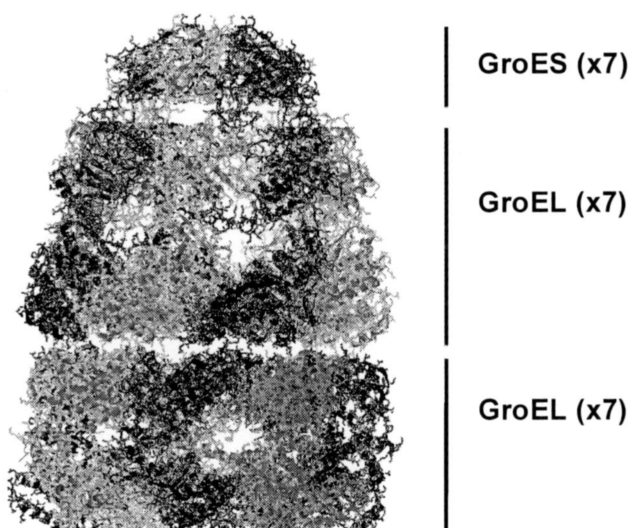

Figure 5. Schematic representation of Hsp60/Hsp10 (GroEL and GroES from *E. coli*). Monomers are black and gray to facilitate visualization. PDB accession number 1PCQ. Figure composed using Weblab ViewerLite 3.7 from Molecular Simulations Inc.

GroEL and GroES (chaperonins) are found in bacteria, mitochondria and chloroplasts, whereas TRiC (TCP-1 ring complex) or CCP (chaperonin containing TCP-1) (where TCP stands for tailless complex polypeptide) are found in the eukaryotic cytoplasmic (Netzer and Hartl, 1998). The mechanism of action of the TCP is not well understood but its structural shape and function are similar to GroEL, i.e. both are double-ring oligomeric complexes. However, TCP is a heterocomplex formed by different units that share about 50% identity and is co-chaperone independent. Although thermosome, an archeal chaperonin, has a built-in lid that seems to work as GroES does. The main function of TCP is to assist the folding of multi-domain proteins in the eukaryote cytoplasm in a co-translational way. TCP can fold the N-terminus of a protein while its C-terminus is being built by the ribosome, which may explain the need for an attached "lid" to keep the cavity open to accommodate the growing polypeptide.

Hsp60 and Hsp10 are essential for plants and the knockout of their genes causes a lethal phenotype in plants (Boston et al., 1996). As a matter of fact, the profile of expressed mRNA belonging to Hsp60 and Hsp10 in sugarcane and eucalyptus are quite interesting (Borges et al, 2001; Cagliari et al., 2005) because proteins belonging to the Hsp60 family are about 15% of all molecular chaperones expressed by sugarcane and almost equally distributed among the three intracellular compartments – mitochondria, chloroplast and cytoplasm. In sugarcane and eucalyptus Hsp60 specifically was found to be expressing twofold of the amount of Hsp10, a result well related to the molecular stoichiometry of 14:7 found in the interaction of these chaperones. Such interaction produces a barrel-like shape structure where protein folding takes place (Figure 5). The expression of cytoplasmic Hsp60 (subfamily TCP-1) is the highest for both sugarcane and eucalyptus, a fact that may be explained because the folding of newly synthesized or imported proteins assistance occurs mainly in the cytoplasm. Members of the Hsp60 family are found in all eucalyptus tissue analyzed but have significant expression in developmental stages which is in good agreement with the fact that *Arabdopsis* species mutated in the Cpn60α of chloroplasts exhibits defects in both embryo and seedlings development.

4.4. HSP90

4.4.1. Struture and Function

Hsp90 is essential for the growth of eukaryotic cells and is a specialized protein folding tool (for a review see Prodromou and Pearl, 2003). Hsp90 is important for maintaining the conformation, stability and function of a wide variety of proteins involved in signal transduction, cell cycle and apoptosis. Hsp90 consists of a highly conserved N-terminal domain (~30 kDa) that contains an ATP-binding site followed by a charged region of unknown function, a flexible domain in the middle (~35 kDa) and a C-terminal domain (~20 kDa) that is responsible for dimerization (Figure 6). Both N- and C-terminal domains have been implicated in substrate binding that is influenced by nucleotide binding.

The function of Hsp90 is driven by a dynamic association with its client proteins and their co-chaperones (HOP, p23, HIP) to form protein complexes. A sub-set of co-chaperones that bind Hsp90 present multiple-copies of a degenerate 34-residue motif which is known as tetratricopeptide repeat (TPR). TPR is formed by a compact helical domain of six consecutive alpha-helices arranged in a regular right-handed superhelix, generating an amphipatic groove that is capable of interacting with 7-12 residues of polypeptide. The binding site for TRP is located in the Hsp90 C-terminus where the domain MEEVD is present (Russell et al., 1999). Hop (Hsc70-Hsp90 organizing protein) recognizes the C-terminus of Hsp90 and Hsp70 and acts as a mediator between them for substrate presentation. The Hsp90 ATPase activity regulates the substrate binding cycle and Hop inhibits this activity by preventing nucleotide binding. The p23 co-chaperone recognizes Hsp90 bound to ATP and stimulates dissociation of the complex. CHIP (*C-terminus of Hsp70-interacting protein*) connect Hsp90 to the ubiquitin complex, helping to control the degradation of proteins by the proteasome.

4.4.2. Hsc70-Hsp90 Chaperone Machinery, an Eukaryotic Protector

Hsc70 and Hsp90 collaborate to form a chaperone machinery, which is a dynamic, multifunctional and multicomponent system that plays a pivotal role in protecting biological organisms from environmental and genetic stresses (Wegele *et al*, 2004). More than 100 proteins (classified mainly in three groups: transcription factors, protein kinases and unrelated proteins) have been identified as substrates for this machinery. In eukaryotes, the Hsc70–Hsp90 machinery is formed by Hsc70, Hsp90, and co-chaperones HOP, Hsp40 and p23. As a matter of fact, it is possible to assemble Hsp90-heterocomplexes with only these five proteins purified. Hsc70 and Hsp90 are sufficient for folding but their co-chaperones modulate, accelerate, potentiate, and stabilize folding. The Hsp70 system has been maintained from prokaryotes to eukaryotes. However, compared to bacteria, the extent of folded polypeptides in the eukaryotic cytosol is more complex and, to handle such diverse conformational requirements, eukaryotes have expanded the central chaperone machinery that was inherited from bacteria. The Hsp90 homologue in bacteria (HtpG) is non-essential and has an accessory function under stress conditions, whereas eukaryotic cytosolic Hsp90 is essential and accomplishes numerous roles. With the exception of Hsp40 homologue, the various co-chaperones of Hsc70 and Hsp90 are not found in bacteria, but arose after the symbiotic development of eukaryotic cells.

The majority of the available experimental data regarding those chaperones are from prokaryotes. Any strategy to deal with human conformational diseases or cancer which involved chaperones will require knowledge about the conformation of these proteins and how they interact with each other. Besides medical and biotechnological importance, the study of this machinery is also relevant to our knowledge of protein interactions inside the cell. For that we have to understand one basic property of chaperones, their ability to specifically target and form non-covalent complexes with other chaperones.

The Hsp90 family represented 10-20% of all 5'EST annotated to molecular chaperone category in eucalyptus and sugarcane, i.e. this family is highly expressed in plant cells in good agreement with the fact that Hsp90 is one of the most abundantly expressed cytoplasmic proteins in several organisms, even in the absence of stress (Boston *et al.*, 1996; Borges *et al.*, 2001). The cytoplasmic Hsp82 is by far the most expressed protein within the Hsp90 chaperone family, an important robservation because Hsp82 is likely to be involved in a net response to low temperature in sugar cane (Nogueira *et al.*, 2003). Hsc70 is also involved in that response to low temperature in sugar cane and it is likely that these two proteins are directly involved in the response to low temperatures in eucalyptus, and for extension to several other plants.

4.4.3. Therapy

Hsp90 is involved in the folding and activation of many proteins involved in apoptosis, signal-transduction pathways and cell-cycle regulation, which are usually deregulated in cancers. As a consequence, Hsp90 has been considered to be an appealing target for cancer therapeutics. Major efforts are leading to many synthesis and biological assays of analogues that resulted in preclinical and Phase I and Phase II studies Sharp and Workman, 2006; Neckers, 2007; and references therein). Therefore the development of Hsp90 inhibitors is considered to be a good example of medicinal chemistry. The number of studies referring to Hsp90 has increased exponentially in the last decade, many of them focused on the syntheses of Hsp90 inhibitors, the development of biological assays to test their effectiveness, and the structure and function of the protein willing to provide more insights on the mode of action of inhibitors and future directions on the medicinal chemistry of Hsp90.

A very important role of Hsp90 is its action as a tumorogenic stabilizer (Sharp and Workman, 2006; Neckers, 2007). Cancer cells exist in a stress situation and several authors have reported induction of members of the Hsc70-Hsp90 chaperone machinery in tumor cells. Several compounds that have been known as antitumor agents by significantly promoting a decrease in the activity oncogenic kinases had Hsp90 as primary target. For instance, Hsp90 has become an interesting target for cancer therapies because it interacts with geldanamycin in tumour cells in a manner that is regulated by Hsp90 co-chaperones and the Hsp70 system. The inhibitors genaldamycin and radicicol bound to the N-terminal domain where the ATP-binding site is located. Due to the nucleotide mimicry of these inhibitors they compete for the same site than ATP. Their affinity for Hsp90 increases 100 fold when this protein is forming heterocomplex with Hsp70, Hsp40, Hop and p23.

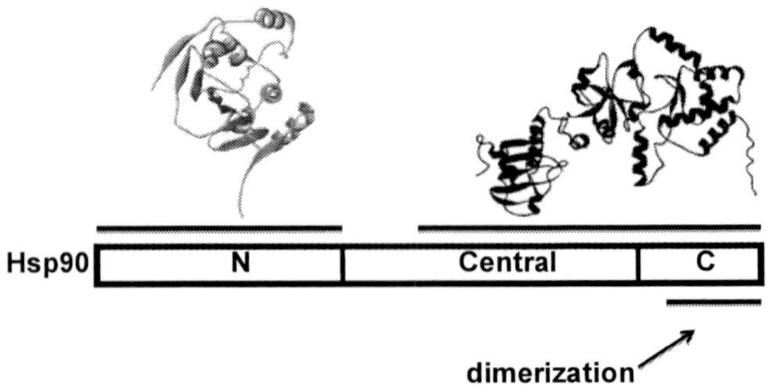

Figure 6. Schematic representation of Hsp90. The N-terminal domain (~30 kDa) contains a nucleotide-binding site, follows a charged region, a central flexible domain (~35 kDa) and a C-terminal domain (~20 kDa) that is responsible for dimerization. Top gray: N-terminal domain residues 2-214 bound to ATP, yeast, PDB accession number 1AM1. Top black: central and C-terminal domains residues 340-735, *Canis familiares*, PDB accession number 2O1T. Figure composed using Weblab ViewerLite 3.7 from Molecular Simulations Inc.

The key role played by Hsp90 in activating many regulatory and signal transduction pathways in the cell has implications for evolution (Whitesell and Lindquist, 2005). Molecular chaperones can help mutated proteins with difficult to fold to reach their correctly native conformation. Without the help of molecular chaperones these deleterious mutations would be negatively selected and disappear. A confirmation for that came from a work of Lindquist´s group showing that inhibition of Hsp90 in *Arabdopsis thaliana* triggered several different phenotypes. Hsp90 is also involved with the biotic stress in plants because it controls the folding of pathogenic proteins. The Hsc70-Hsp90 machinery also plays an important role during infection by microorganisms because it is recruited to facilitate the formation of a complex comprised of viral protein and RNA. As one easily concludes this machinery has high biotechnological and medical importance.

4.5. HSP100

4.5.1. Struture and Function

The proteins of the Hsp100 family have a molecular mass ranging from 80 to 105 kDa, have ATPase activity and form ring oligomers. Proteins from the Hsp100 family are involved in several biological phenomena, including stress tolerance, proteolysis, DNA transposition and gene regulation. Hsp100 is essential for thermotolerance because its deletion goes unnoticeable to the cell at normal conditions but is deleterious at heat shock. *Sacharomicies cerevisiae* Hsp100, Hsp104, has 102 kDa and contains two binding sites for ATP and ATP hydrolysis is important for the translocation of polypeptide substrate through a central channel formed in the oligomeric assembly of Hsp100. The model for Hsp100 based on structural information is that of a propeller formed by 6 monomers with a central cavity (Figure 7). This hexameric form, which is active and stabilized by nucleotides, is probably in equilibrium with other species (monomer, dimer and/or trimer) which are inactive (Parsell et al, 1994). The binding of ATP would then triggers the oligomerization of Hsp100 and the energy provided by the hydrolysis of ATP would drive the translocation of polypeptides through the central channel.

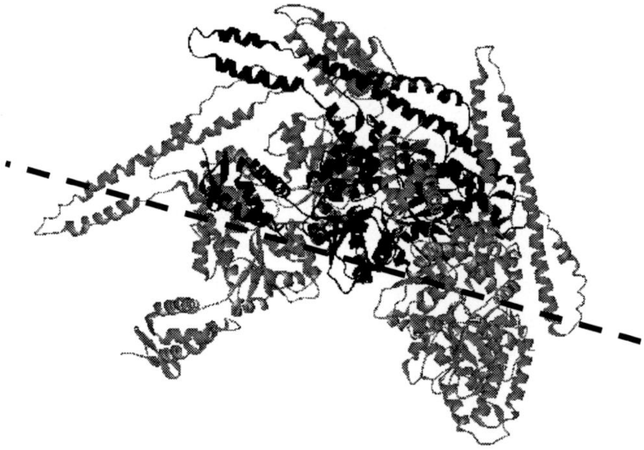

Figure 7. Schematic representation of the sctruture of Hsp100. Structure of *Thermus thrmophilos* ClpB, PDB accession number 1QVR. The structure is that of a trimer (two subunits in gray and one in black). A second trimer will be located at the other side of the dashed line forming a hexamer in which six helices will be pointing away from the propeller and the central part will be forming a channel. Figure composed using Weblab ViewerLite 3.7 from Molecular Simulations Inc.

The power to dissolubilize aggregates may be one of the best finds for chaperones because it opened the possibility to use them as a therapy against diseases caused by misfolded proteins involved in diseases. One of the first indications of this capability was that studied by Prof. Bukau's group where they investigated the effect of chaperones on the thermostability of *Escherichia coli* (Mogk et al., 1999').The authors used knocked-out mutants of *E. coli* in which ClpB, Hsp100 homologue, was deleted and analyzed cell standardly and heated-shock treated cells by two dimension-gel electrophoresis. They were able to show an increasing in protein aggregation and found that molecular chaperones belonging to the Hsp100 family arekey players in this process (see below). However further results showed that Hsp100 was not merely involved in the suppression of aggregation. Using unfolded luciferase as a model, Glover and Lindquist (1998) showed that Hsp100 is chiefly involved in the disaggregation of protein aggregates formed during heat stress. Such aggregates are sometimes not protect by the other chaperones due to overload capacity.

4.5.2. Therapy

The ability of DnaK and ClpB of dissolving protein aggregates was tested by Mogk et al. (1999). Soluble *E. coli* extracts were pre-incubated at 30°C for 5min in the absence and in the presence of chaperones and ATP was added 2 min prior heat-shock. Then samples were shifted to 45°C for 15min and aggregated proteins were pelleted by centrifugation and analyzed by one dimension- and two dimension-gel electrophoresis. The protein aggregates were isolated by centrifugation and ressuspended in buffer with or without combination of chaperones and ATP and incubated at 30°C for 4hs. Solubilization of aggregates in the order of 3-5% was observed with DnaK alone and up to 50% when both DnaK and ClpB were added (1:1 monomer stoichiometry, 2 micromolar). Two other important observations were that the disaggregation strictly depended on ATP and protein aggregation in ΔdnaK was reversed by the addition of ClpB and DnaK. These findings confirmed a stress-related role of the cooperative activity of the DnaK system and ClpB. The use of a specific ClpB inhibitor was also tested and did not erased completely the function of DnaK in disaggregating and refolding some aggregates *in vitro* suggesting that DnaK alone has the ability to refolding some substrates. An additional result, important to the understanding of the complexity of molecular chaperones in the cell, was that the addition of GroEL/GroES significantly activated the refolding. These findings led to a proposal for a sequencing action of chaperones to disaggregate and refolding aggregates: ClpB processes compact aggregates resulting in loose aggregates that can be disaggregated by DnaK to partially folded chains which will reach native conformation with the help of GroEL/GroES. (Ben-Zvi and Goloubinoff, 2001).

Much work has also been done with Hsp104, the Hsp100 from *S. cerevisiae* (Glover and Lindquist, 1998). We now know: 1) The deletion of the *hsp104* gene in yeast stopped the recovering of aggregated proteins in cells exposed to heat shock conditions. Additional experiments were confirmatory that the ressolubilization of Hsp104 was only possible in the presence of both Ssa1 and Ydj1. 2) Ssa1 and Ydj1 are factors that co-eluted with Hsp104 in chromatography experiments and GroEL and Hsp70 are able to suppress the aggregation of luciferase. 3) Yeast Hsp104 is not able to prevent the aggregation of luciferase but is able to stimulate, in an ATP-dependent manner, the reactivation of luciferase in the substrate from an *hsp104* mutant cell. 4) SmHsps seem also to be part of the multi-chaperone network that cooperates with Hsp100. 5) The maintenance and inheritance of the aggregated form of Sup35 in vivo, a protein that undergoes self-seeded assembly into amyloid fibrils in vitro, requires Hsp104. And 6) Hsp104 also promotes the *in vitro* conversion of PrP to an aggregated and protease resistant form similar to PrP^{Sc}.

4.6. SmHSP

4.6.1. Struture and Function

Proteins belonging to the small heat shock protein (SmHsp) family are ubiquitous in all types of organisms, including archaea, bacteria, and eukarya. The importance of SmHsp is showed by the fact that the main heat shock proteins detected in plants are proteins with 15-28 kDa (Key et al., 1981). They accumulate up to 1% in leafs, are correlated with the acquisition of thermotolerance and bind partially denatured proteins preventing irreversible protein aggregation and inactivation. In this sense SmHsps are 'holdases' and SmHsp-bound proteins are effective substrates for refolding by other chaperones. Transgenic plants that have enhanced expression of SmHsps are tolerant to high temperatures.

Small Hsps are usually found in reproductive organs during several stages of plant development and after heat (or other) shock induction which is in agreement with the high levels of Small Hsp 5'ESTs found in libraries originated from eucalyptus stem tissues susceptible to hydric deficit (Cagliari et al., 2005). Similar results were found for other plants such as sunflowers and *A. thaliana*, in which an increase in small Hsp expression correlates with osmotic stress (Sun *et al.*, 2002). Also, the dehydration can contribute to aggregation caused by the increasing concentration of proteins inside the cells.

SmHsps have a highly variable N-terminus and are grouped together based on a conserved domain named α-crystallin after the α-crystallin of the vertebrate eye lens (Caspers et al., 1995) (Figure 8A). The N-terminus is implicated with substrate interaction: 1) the N-terminus of SmHsp26 of *S. cerevisiae* is involved with both the assembly and the interaction with non-native proteins (Stromer et al., 2004); 2) studies using a cryo-electron microscopy tool showed that this protein experience a temperature-induced conformational change in which substrate binding sites located at unstructured regions of the N-terminus are exposed (White et al. (2006): and 3) two sugarcane SmHsps which differ in structure at a specific domain in the N-terminus have different substrate specificity (Tiroli and Ramos, 2007).

The two sugarcane SmHsp chaperones studied are the most expressed genes belonging to the SmHsp family in sugarcane (Borges et al., 2001) and they may be acting upon different substrates during heat shock. There are two hypotheses for the mechanisms of action of SmHsp. In one of them the oligomer exposes binding sites to unfolded proteins upon heat shock and in the other the oligomer dissociates in smaller units that expose binding sites to unfolded proteins upon heat shock (Figure 8B). Much work is in progress to determine whether which one or both hypotheses are true.

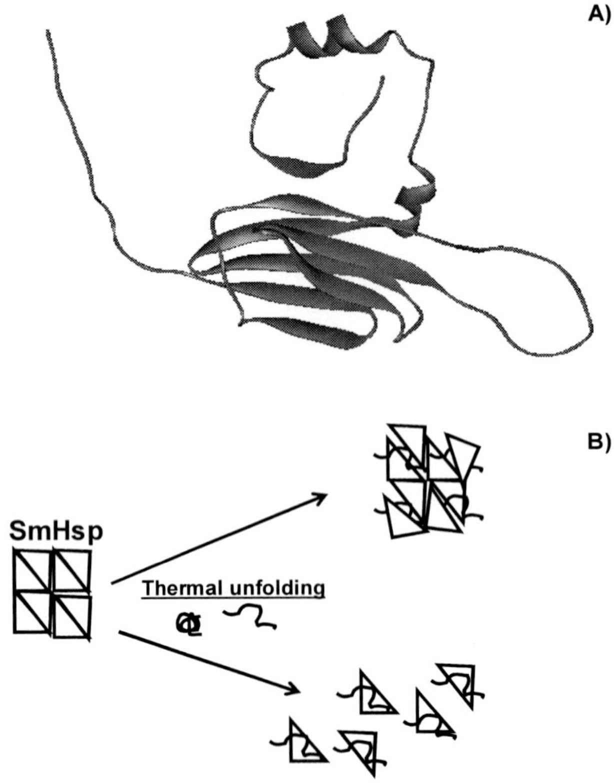

Figure 8. A) Structure of the folded monomer of smHsp16.9 form wheat, PDB accession number 1GME. Figure composed using Weblab ViewerLite 3.7 from Molecular Simulations Inc. B) Schematic representation of two possible mechanisms of action of SmHsp. Top: the oligomer exposes binding sites to unfolded proteins upon heat shock. Bottom: the oligomer dissociates in smaller units that expose binding sites to unfolded proteins upon heat shock.

4.6.2. Therapy

Mutations in SmHsps are identified as responsible for several human pathologies such as human cardiovascular disorder (Desmin Related Myophaty) and hereditary cataract. Lens tissue is formed by proteins known as crystallin which are essential for vision and are not renewed. It was shown that crystallins in solution have reduced solubility with time, which leads to aggregation and formation of fibrils, a mechanism considered similar to that causing cataract. As a result, perturbations in the chaperone activity can lead to decrease in the transparency of lens.

SmHsps are highly expressed in several neurodegenerative disorders and aggregation is inhibited when sHsp is upregulated (Clark et al., 2000). Proteins belonging to both the Hsp70 and SmHsp families cominunoprecipitate with the β-amiloid peptide in a transgenic *Caenorhabditis elegans* Alzheimer model created by Link´s group (Fonte et al., 2002). In this model, the β-amiloid peptide accumulates in intracellular cytoplasmic deposits in a similar way to that present in Alzheimer diseases. In addition to show that SmHsps bind to the β-amiloid peptide, they also showed that the expression of SmHsps is induced by the expression of the peptide. Furthermore, enhanced expression of a specific SmHsp, Hsp16.2 is able to suppress the deleterious effect of the peptide *in vivo*. SmHsps seem also to be involved in cancer since specific proteins from this family can be used as targets in antigen tests and superexpression of human Hsp27 in melanoma cells slows growth inhibiting tumor progression. To sum up, SmHsps are becoming potential targets for therapeutic approaches once their structure-function relationship is established.

5. Conclusion

Although proteins can fold *in vitro* without the aid of additional components they seem to need assistance of molecular chaperones to fold to their native state *in vivo*. The diseases linked to the failure of proteins to reach their native state are usually known as misfolding diseases and their development are influenced by molecular chaperones. Molecular chaperones act by preventing protein folding and aggregation and by disaggregating misfolded proteins. Therefore, the functional failure of the molecular chaperone machinery is a potential trigger of degenerative conditions. The actions of molecular chaperones are so crucial to the cell that they have been extensively studied as potential targets for therapeutic approaches aimed in decreasing the toxicity of aggregates. It remains to be seen whether therapies involving molecular chaperones will be successful in reverting misfolded diseases, however a better understanding on the mechanism of action of molecular chaperones will truly contribute to provide insights into cellular function at normal and stressed conditions.

6. References

Acebrón S.P., Fernández-Sáiz V., Taneva S.G., Moro F., Muga A. (2008) DnaJ recruits DnaK to protein aggregates. *J Biol Chem. 283*, 1381-1390.

Anfinsen C.B., Haber E., Sela M., White F.H. Jr. (1961) The kinetics of formation of native ribonuclease during oxidation of the reduced polypeptide chain. *Proc Natl Acad Sci U S A. 47*, 1309-1314.

Beissinger, M., Buchner, J. (1998) How chaperones fold proteins. *Biol. Chem., 379*, 245-259.

Ben-Zvi, A.P. Goloubinoff, P. (2001) Review: mechanisms of disaggregation and refolding of stable protein aggregates by molecular chaperones. *J. Struct. Biol. 135*, 84-93.

Bonini N.M. (2002) Chaperoning brain degeneration. *Proc Natl Acad Sci U S A. 99*, 16407-16411.

Borges, J.C., Peroto, M.C., Ramos, C.H.I. (2001). Molecular Chaperone Genes in the SugarCane Expressed Sequence Database (SUCEST). *Gen. Mol. Biol. 24,* 85-92.

Borges, J.C., Hannes, F., Craievich, A.F., Ramos, C.H.I. (2005). Low-resolution structural study of two human Hsp40 chaperones in solution. HJA1 from subfamily A and HJB4 from subfamily B, have different quaternary structures. *J. Biol. Chem. 280*, 13671-13681.

Borges, J.C., Ramos, C.H.I. (2005) Protein folding assisted by chaperones. *Protein Pept. Lett., 12*, 256-261.

Borges, J.C., Ramos, C.H.I. (2006) Spectroscopic and thermodynamic measurements of nucleotide-induced changes in the human 70-kDa heat shock cognate protein. *Arch. Biochem. Biophys. 452*, 46-54.

Boston, R.S., Viitanen, P.V., Vierling, E. (1996) Molecular chaperones and protein folding in plants. *Plant Mol. Biol., 32*, 191-222.

Cagliari, T.C., Tiroli, A.O., Borges, J.C., Ramos, C.H.I. (2005) Identification and in silico expression analysis of eucalyptus expressed sequencing tags (ESTs) encoding molecular chaperones. *Gen. Mol. Biol. 28*, 520-528.

Caspers, G.J., Leunissen, J.A., de Jong, W.W. (1995) The expanding smallheat shock protein family, and structure predictions of the conserved"alpha-crystallin domain." *J. Mol. Evol. 40*, 238–248.

Chaduri, T.K., Paul, S. (2006) Protein-misfolding diseases and chaperone-base therapeutic approaches. *The FEBS J. 273*, 1331-1349.

Cheetham, M.E., Caplan, A.J. (1998) Structure, function and evolutionof DnaJ: Conservation and adaption of chaperone function. *Cell Stress Chap. 3*, 28–36.

Clark, J.I., Muchowski, P. (2000) Small heat-shock proteins and their potential role in human disease. *Curr. Opin. Struct. Biol., 10,* 52-59

Craig, E.A. (1989) Essential roles of 70kDa heat inducible proteins. *Bioessays, 11*, 48-52.

Csermely, P. (2001) Chaperone overload is a possible contributor to 'civilization diseases'. *Trends Genet. 17*, 701-704.

Cyr, D.M., Langer, T., Douglas, M.G. (1994) DnaJ-like proteins: molecular chaperones and specific regulators of Hsp70. *Trends Biochem. Sci. 19*, 176–181.

Daugaard, M., Rohde, M., Jäättelä, M. (2007) The heat shock protein 70 family: highly homologous proteins with overlapping and distinct functions. *FEBS Lett., 581*, 3702-10

Dobson, C.M. (2003) Protein folding and misfolding. *Nature. 426*, 884-90.

Dobson, C.M. (2004) Experimental investigation of protein folding and misfolding. *Methods 34*, 4-14.

Dyson, H.J., Wright, P.E.. (2005) Intrinsically unstructured proteins and their functions. *Nat Rev Mol Cell Biol. 6*, 197-208.

Ellis, R.J., Hartl F.-U. (1996) Protein folding in the cell: competing models of chaperonin function. *FASEB J. 10*, 20-26.

Ewalt, K.L., Hendrick, J.P., Houry, W.A., Hartl, F.-U. (1997) *In vivo* observation of polypeptide flux through the bacterial chaperonin system. *Cell. 90*, 491-500.

Fan, C.Y., Lee, S., Re, H.Y., Cyr, D.M. (2004) Type I and type II Hsp40s containexchangeable chaperone modules that specify Hsp70 function. *Mol Biol Cell, 15*, 761-73.

Fandrich, M., Fletcher, M.A., Dobson, C.M. (2001) Amyloid fibrils from muscle myoglobin. *Nature 410*, 165-166.

Fandrich, M., Dobson, C.M. (2002) The behaviour of polyamino acids reveals an inverse side chain effect in amyloid structure formation. *EMBO J. 21*, 5682-5690.

Fonte, V., Kapulkin, V., Taft, A., Fluet, A., Friedman, D., Link, C.D. (2002) Interaction of intracellular beta amyloid peptide with chaperone proteins. *Proc Natl Acad Sci U S A. 99*, 9439-9444.

Georgopoulos, C.P., Hendrix, R.W., Kaiser, A.D., Wood, W.B. (1972) Role of the host cell in bacteriophage morphogenesis: effects of a bacterial mutation on T4 head assembly. *Nat New Biol. 239*, 38-41.

Glover, J.R., Lindquist, S. (1998) Hsp104, Hsp70, and Hsp40: a novel chaperone system that rescues previously aggregated proteins. *Cell. 94*, 73-82.

Goloubinoff, P., Christeller, J.T., Gatenby, A.A., Lorimer, G.H. (1989) Reconstitution of active dimeric ribulose bisphosphate carboxylase from an unfoleded state depends on two chaperonin proteins and Mg-ATP. *Nature. 342*, 884-889.

Hartl, F.-U., Hayer-Hartl, M. (2002) Molecular chaperones in the cytosol, from nascent chain to folded protein. *Science, 295*, 1852-1858.

Hendrick, J.P., Hartl, F.-U. (1993) *Annu. Rev. Biochem., 62*, 349-84.

Jiang, J., Prasad, K., Lafer, E.M., Sousa, R. (2005) Structural basis of Interdomain Communication in the Hsc70 Chaperone. *Mol. Cell., 20*, 513-524.

Kaiser, C.M., Chang, H.C., Agashe, V.R., Lakshmipathy, S.K., Etchells, S.A., Hayer-Hartl M., Hartl, F.-U., Barral, J.M. (2006) Real-time observation of trigger factor function on translating ribosomes. *Nature. 444*, 455-460.

Key, J.L., Lin, C.Y., Chen, Y.M. (1981). Heat shock proteins of higher plants. *Proc. Natl. Acad. Sci., 78,* 3526-3530.

Laskey, R.A., Honda, B.M., Mills, A.D., Finch, J.T. (1978) Nucleosomes are assembled by an acidic protein which binds histones and transfers them to DNA. *Nature, 275*, 416-420.

Lian, H.Y., Zhang, H., Zhang, Z.R., Loovers, H.M., Jones, G.W., Rowling, P.J., Itzhaki, L.S., Zhou, J.M., Perrett, S. (2007) Hsp40 interacts directly with the native state of the yeast prion protein Ure2 and inhibits formation of amyloid-like fibrils. *J Biol Chem. 282*, 11931-11940.

Liberek, K., Marszalek, J., Ang, D., Georgopoulos, C., Zylicz, M. (1991) Escherichia coli DnaJ and GrpE heat shock proteins jointly stimulate ATPase activity of DnaK. *Proc Natl Acad Sci 88*, 2874–2878.

Lopez, N., Aron, R., Craig, E.A. (2003) Specificity of class II Hsp40 Sis1in maintenance of yeast prion [RNQ(1)]. *Mol Biol Cell 14*, 1172–1181.

Lorimer, G.H. (2001) A personal account of chaperonin history. *Plant Physiol. 125*, 38-41.

Lu, Z., Cyr, D.M. (1998). Protein folding activity of Hsp70 is modified differentially by the hsp40 co-chaperones Sis1 and Ydj1. *J Biol Chem 273*, 27824–27830.

Macario, A.J.L. de Macario, E.C. (2007). Chaperonophaties and chaperonotherapy. *FEBS Lett. 581*, 3681-3688.

Mayer, M.P., Brehmer, D., Gässler C.S., Bukau, B. (2001) Hsp70 Chaperone Machines. *Adv. Protein Chem., 50*, 1-45.

Minton, A.P. (2000). Implications of macromolecular crowding for protein assembly. *Curr. Op. Struct. Biol. 10*, 34-39.

Moriyama, H.,, Edskes, H.K.,Wickner, R.B. (2000) [URE3] prion propagation in Saccharomyces cerevisiae: requirement for chaperone Hsp104 and curing by overexpressed chaperone Ydj1p. *Mol Cell Biol 20*, 8916-8922

Neckers, L. (2007) Heat shock protein 90: the cancer chaperone. *J Biosci 32*, 517-530Netzer, W.J., Hartl, F.-U. (1998) Protein folding in the cytosol: chaperonin-dependent and -independent mechanisms. *Trends Biochem Sci. 23*, 68-73.

Nogueira, F.T., De Rosa, V.E.Jr, Menossi, M., Ulian, E.C., Arruda, P. (2003) RNA expression profiles and data mining of sugarcane response to low temperature. *Plant Physiol. 132*, 1811-1824.

Parsell, D.A., Kowal, A.S., Lindquist, S. (1994) *Saccharomyces cerevisiae* Hsp104 protein. Purification and characterization of ATP-induced structural changes. *J Biol Chem. 269*, 4480-4487.

Prodromou, C., Pearl, L.H., (2003). Structure and functional relationships of Hsp90. *Curr Cancer Drug Targets. 3*, 301-23.

Prusiner, S.B. (2001) Shattuck lecture--neurodegenerative diseases and prions. *N Engl J Med. 344*, 1516-1526.

Ramos, C.H.I., Ferreira, S.T. (2005). Protein folding, misfolding and aggregation: evolving concepts and conformational diseases. *Protein Pept. Lett. 12*, 213-222.

Russell, L.C., Whitt, S.R., Chen, M.S., Chinkers, M. (1999) Identification of conserved residues required for the binding of a tetratricopeptide repeat domain to heat shock protein 90. *J Biol Chem. 274*, 20060-20063

Schwimmer, C., Masison, D.C. (2002) Antagonistic interactions between yeast [PSI(+)] and [URE3] prions and curing of [URE3] by Hsp70 protein chaperone Ssa1p but not by Ssa2p. *Mol Cell Biol 22*, 3590-3598

Sharp S, Workman P. (2006) Inhibitors of the HSP90 molecular chaperone: current status. *Adv Cancer Res 95*, 323-348

Stromer, T, Fischer, E, Richter, K, Haslbeck, M, Buchner, J, (2004) Analysis of the regulation of the molecular chaperone Hsp26 by temperature-induced dissociation, *J. Biol. Chem. 279*, 11222-11228.

Teter, S.A., Houry, W.A., Ang, D., Tradler, T., Rockabrand, D., Fischer, G., Blum, P., Georgopoulos, C., Hartl, F.-U. (1999). Polypeptide Flux through Bacterial Hsp70DnaK Cooperates with Trigger Factor in Chaperoning Nascent Chains. *Cell, 97*, 755-765

Theyssen, H., Schuster, H-P., Packschies, L., Bukau, B., Reinstein, J. (1996) The Second Step of ATP Binding Induces Peptide Release. *J. Mol. Biol. 263*, 657-670.

Tiroli, A.O., Ramos, C.H.I. (2007). Biochemical and biophysical characterization of small heat shock proteins from sugarcane. Involvement of a specific region located at the N-terminus with substrate specificity. *Int. J. Biochem. Cell Biol. 39*, 818-831.

Wegele, H., Müller, L., Buchner, J. (2004) Hsp70 and Hsp90 – a relay team for protein folding. *Rev. Physiol. Biochem. Pharmacol., 151*, 1-44.

White, H.E., Orlova, E.V., Chen, S., Wang, L., Ignatiou, A., Gowen, B. Stromer, T., Franzmann, T.M., Haslbeck, M., Buchner, J., Saibil, H.R. (2006) Multiple distinct assemblies reveal conformational flexibility in the small heat shock protein hsp26. *Structure 14*, 1197-1204.

Whitesell, L., Lindquist, S.L (2005) HSP90 and the chaperoning of cancer. *Nat Rev Cancer 10,* 761-772.

INDEX

3

A

B

C

D

E

F

G

H

I

J

N

O

P

Q

R

S

T

B